MATHS PROBLEM SOLVING
RESULTS BOOSTER

11+

11+
PROBLEM SOLVING
FOR CEM
RESULTS BOOSTER

FAISAL NASIM

Contents

Guidance for Parents

About the 11+ tests

In most cases, the 11+ selection tests are set by GL Assessment (NFER), CEM (The University of Durham) or the individual school. You should be able to find out which tests your child will be taking on the website of the school they are applying to, or from the local authority.

The CEM test usually consists of two papers and in each paper pupils are tested on their abilities in verbal, non-verbal and numerical reasoning. Tests are separated into small, timed sections accompanied by audio instructions. It appears the content required for CEM tests changes from year to year, and the level at which pupils are tested ranges from county to county.

For pupils to do well in the CEM tests:

- they must have strong arithmetic skills
- they must have strong reasoning and problem-solving skills
- they must have a strong core vocabulary
- they must be flexible and able to understand and respond to a wide range of question types and formats, without being panicked by unfamiliar question types
- they must be able to work under time pressure.

This book provides preparation for the problem solving part of the exam and, more specifically, the multi-step problems. These tests will help students to develop the problem solving fundamentals that will allow them to work quickly and accurately in the actual exam.

Helpful tips are given for questions where students commonly make mistakes – encourage your child to follow these tips to help ensure they don't lose vital marks in the exam.

Your child should **not** use a calculator for any of these tests.

The importance of practice

Practice will help your child to do his or her best on the day of the tests. Working through the more difficult question types allows your child to practise answering a range of test-style questions that will help them achieve the best results. It also provides an opportunity for them to learn how to manage time effectively, so that time is not wasted during the test and any 'extra' time is used constructively for checking.

How to prepare for the 11+ tests

- Use this book to help your child get used to answering questions under time constraints.
- Help your child revise and practise problem areas.
- Help your child gain confidence in their abilities.
- Talk about coping with pressure.
- Let your child know that tests are just one part of school life.
- Let them know that doing their best is what matters.
- Plan a fun incentive for after the 11+ tests, such as a day out.

Test 1: Algebra

INSTRUCTIONS

You have 10 minutes to complete the following section.
You have 10 questions to complete within the time given.

Circle the letter beside the correct answer.

EXAMPLE

$2x + 1 = 9$
What is the value of x?

(A) 4
B 5
C 6
D 7
E 8

The answer is **A**.
$2x = 9 - 1 = 8; x = 8 \div 2 = 4$

(1) Gabby had $4n$ pencils and her sister had half as many. They divided the total number of pencils equally between the 2 of them. How many pencils does Gabby have now?

A $2n$
B $4n$
C $3n$
D $7n$
E $\frac{3}{2}$

(2) An isosceles triangle is shown here. If the perimeter of the triangle is 40.4 cm, what is the value of x?

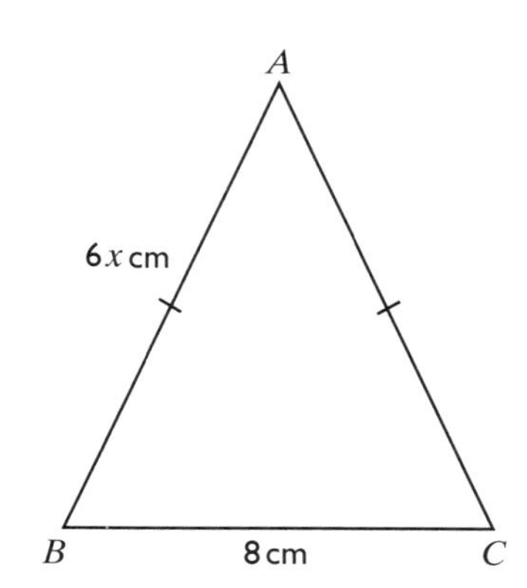

A 2.1
B 32.4
C 2.6
D 2.7
E 1.8

3. 12 years ago, Mandy was s times as old as Debbie was. If Debbie is 40 now, how old is Mandy now?

A $28s + 12$
B $40s - 12$
C $28s - 12$
D $40s + 12$
E $32s - 12$

4. The price of a pizza and 4 drinks is equally split between 4 friends. If each of them paid t and the 4 drinks cost a total of d, what was the cost of the pizza?

A $4t + d$
B $\frac{4t}{d}$
C $\frac{t}{4} - d$
D $4t - d$
E $\frac{t}{4} + d$

5. If $a = 4$ and $b = -5$, what is the value of $\frac{b^2}{2a}$?

A 3.25
B 3.125
C 6.25
D −3.125
E 8.25

6. Figure B shows the intersection of 2 straight lines. What is the value of $(p + 7)$?

A 10
B 13
C 20
D 65
E 21

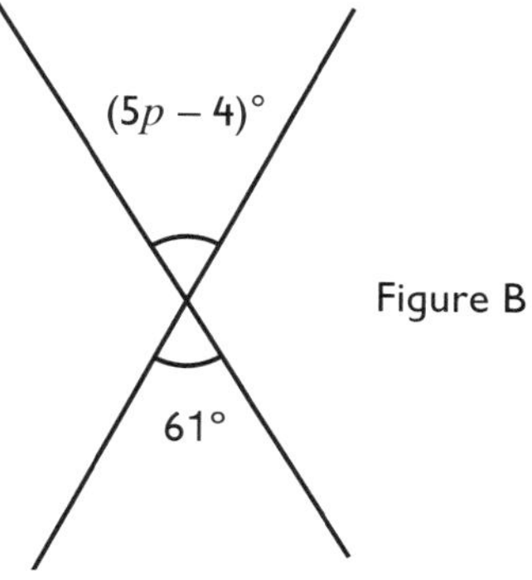

Figure B

7. The ages of Abi and Bella are in the ratio 3:5. In 9 years, their ages will be in the ratio 3:4. How old is Bella now?

A 10
B 15
C 12
D 20
E 5

Questions continue on next page

8) This table shows the number of songs downloaded by Sally over 5 days.

Day	Cost in pence per song	Number of songs
Monday	c	12
Tuesday	d	15
Wednesday	c	5
Thursday	c	10
Friday	d	8

Which expression shows the total amount of money she spent downloading songs over these 5 days?

A £$(27c + 23d)$

B $\frac{£(27c + 23d)}{100}$

C $\frac{£(50cd)}{100}$

D £$50cd$

E £$\left(27c + \frac{23d}{100}\right)$

Use the diagram below to help you answer questions 9 and 10.

The jug to the right is $\frac{3}{8}$ full of water, as indicated by x.

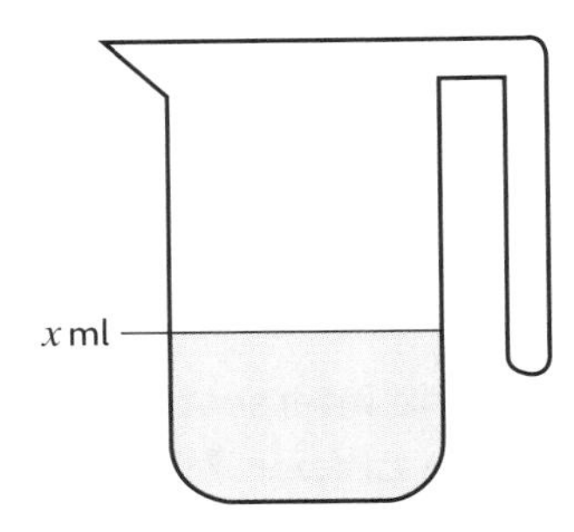

9) What is the capacity of the jug in terms of x?

A $\frac{8x}{3}$ ml

B $24x$ ml

C $11x$ ml

D $\frac{3x}{8}$ ml

E $8x + 3$ ml

10) Leon drinks 30 ml of water from the jug. Which expression shows how much water remains in the jug in ml?

A $30 - x$

B $3x - 90$

C $\frac{(x - 30)}{8x}$

D $x - 30$

E $30x - \frac{3}{8}$

Test 2: Decimals

INSTRUCTIONS

You have 10 minutes to complete the following section.
You have 10 questions to complete within the time given.

Circle the letter beside the correct answer.

EXAMPLE

What is 0.5 greater than 45.7?

A 46.3
(B) 46.2
C 46.5
D 47.2
E 44.2

The answer is **B**.
45.7 + 0.5 = 46.2

1. William divides X by 10. Sam divides X by 0.1. Which of the following statements is correct?

A Sam's answer is 100 times greater than William's answer.
B Sam's answer is 10 times greater than William's answer.
C Sam's answer is 10 times smaller than William's answer.
D Sam's answer is 100 times smaller than William's answer.
E Sam's answer is 0.1 times smaller than William's answer.

2. Andrea multiplies 2 of the numbers shown below and then rounds her answer to the nearest whole number. Her answer is 47. Which 2 numbers did she multiply?

1.2	36	1.5	24	1.96
V	**W**	**X**	**Y**	**Z**

A V and W
B W and X
C Y and Z
D X and Y
E V and Y

Questions continue on next page

(3) In 1 day, Mary uses 10.1 units of electricity and 15.2 units of gas. This table shows the cost of electricity and gas per unit.

Electricity	Gas
20 p	10 p

If Mary receives a 50% reduction off her bill, what is the total amount she must pay for Electricity and Gas usage on this day?

A £1.50
B £3.54
C £1.77
D 175 p
E £0.90

(4) A school orders 2,000 pencils that cost £0.21 each. The school has a total budget of £3,000 for stationery. What percentage of this stationery budget is used to buy the pencils?

A 15%
B 12%
C 14%
D 4%
E 20%

(5) Fernando wishes to check in some baggage for a flight. He has 2 bags that weigh 12.9 kg and 7.5 kg. He also has 5 souvenirs, each weighing 600 g. The airline allows a total of 23 kg of baggage per passenger. By how much weight does he exceed the baggage allowance of the airline?

A 300 g
B 600 g
C 220 g
D 400 g
E 500 g

(6) Raya's coffee estate produces 120.4 kg of coffee. She packs the coffee into bags weighing 200 g each. She then sells $\frac{1}{7}$ of these bags for £3.50 each. How much money does she receive from the sale?

A £300
B £301
C £380
D £2,902
E £2,107

(7) 2 boxes of chocolates weigh as much as 4 boxes of sweets. If the total weight of all 6 boxes is 8.8 kg, what is the weight of each box of sweets?

A 110 g
B 1.1 kg
C 200 g
D 900 g
E 800 g

(8) The odometer to the right shows the number of miles travelled by Eric on a trip that lasted 6.5 days.

8	7	6	.	2

What is the average distance Eric covered per day?

Round your answer to the nearest mile.

A 135 miles
B 120 miles
C 520 miles
D 156 miles
E 160 miles

(9) This table shows the rainfall in London over 4 months.

	Rainfall
November	10.2 cm
December	8.6 cm
January	11.56 cm
February	?

If the mean rainfall over these 4 months was 9.59 cm, what was the rainfall in February?

A 15 cm
B 4 cm
C 8 cm
D 10 cm
E 6 cm

(10) 320 identical candies are divided into 2 piles, as shown to the right.

A	B
200 candies	120 candies
4.4 kg	?

What is the total weight of the candies in Pile B?

A 2.2 kg
B 2.8 kg
C 1.8 kg
D 2.28 kg
E 2.64 kg

Test 3: Fractions

INSTRUCTIONS

You have 10 minutes to complete the following section.
You have 10 questions to complete within the time given.

Write the correct answer in the boxes provided (one digit per box).

EXAMPLE

Ben thinks of a number. Half of the number is 8 greater than one third of the number.

What number is Ben thinking of? The answer is:

Let the number be X; $\frac{1}{2}X - \frac{1}{3}X = 8$

$\frac{3}{6}X - \frac{2}{6}X = 8$; $\frac{1}{6}X = 8$; $X = 6 \times 8 = \mathbf{48}$

1. A spelling test has 25 questions. To be awarded Grade A, at least $\frac{3}{4}$ of the questions must be answered correctly. The scores of a group of 5 friends are shown below.

Name	Marks out of 25
Carl	18
Violet	21
Claire	16
Emily	21
Nikki	19

What percentage of the group was awarded Grade A?

%

2. A baby whale weighs 49 kg in March. Its weight increases by $\frac{2}{5}$ over the next 3 months. What is the weight of the baby whale after 3 months?

☐☐.☐ kg

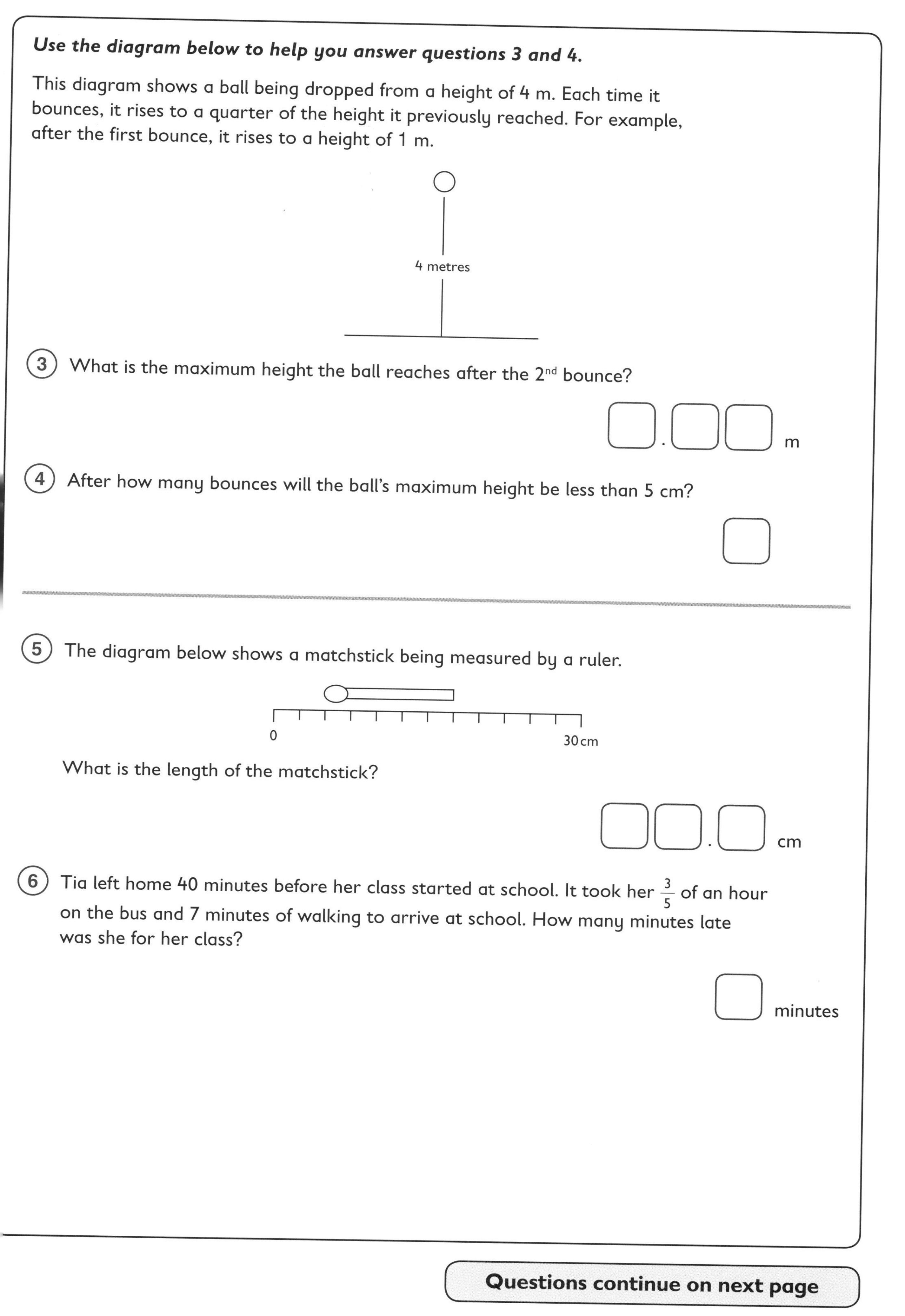

Use the diagram below to help you answer questions 3 and 4.

This diagram shows a ball being dropped from a height of 4 m. Each time it bounces, it rises to a quarter of the height it previously reached. For example, after the first bounce, it rises to a height of 1 m.

(3) What is the maximum height the ball reaches after the 2nd bounce?

☐.☐☐ m

(4) After how many bounces will the ball's maximum height be less than 5 cm?

☐

(5) The diagram below shows a matchstick being measured by a ruler.

What is the length of the matchstick?

☐☐.☐ cm

(6) Tia left home 40 minutes before her class started at school. It took her $\frac{3}{5}$ of an hour on the bus and 7 minutes of walking to arrive at school. How many minutes late was she for her class?

☐ minutes

Questions continue on next page

Use the diagram below to help you answer questions 7 and 8.

This diagram shows Shape A, which is constructed from identical cube blocks.

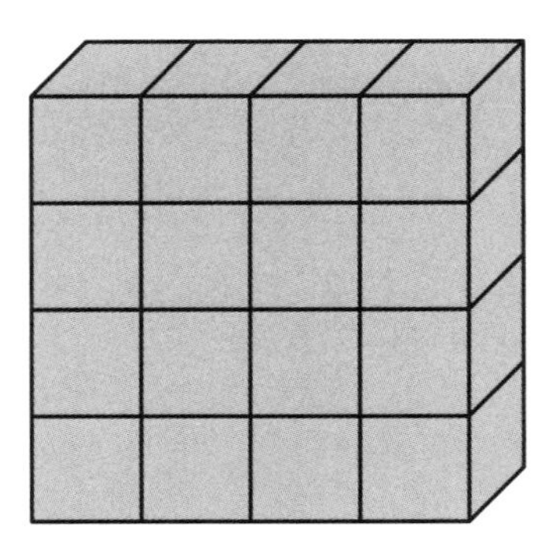

(7) Jose paints the faces of the blocks in Shape A that are surrounded by other blocks on four of their sides. What fraction of Shape A's surface does he paint?

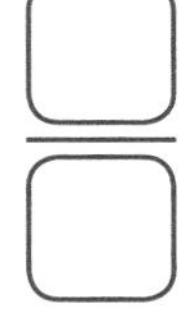

(8) Jose stacks 2 more identical shapes side by side with Shape A to make a new shape. $\frac{1}{8}$ of the blocks in the new shape are made of steel. How many blocks are made of steel?

---

(9) In an election, $\frac{2}{5}$ of the voters voted for Mr Yew and $\frac{7}{20}$ voted for Miss Garswood. The rest voted for Mr Widmore. What percentage of voters voted for Mr Widmore?

 %

HELPFUL TIP: *Think about what calculation you can use to turn fractions into percentages.*

(10) Lucy baked 322 cupcakes. She sold $\frac{2}{7}$ of them for 60 p each. She also baked 200 cookies and sold half of them for 90 p each. She spent £80 to set up her stall and pay for ingredients. After subtracting the money that she spent, how much of the money that she made by selling the cupcakes and cookies remained?

£ ☐☐.☐☐

Test 4: Angles

INSTRUCTIONS

You have 10 minutes to complete the following section.
You have 10 questions to complete within the time given.

Circle the letter beside the correct answer.

EXAMPLE

How much greater is the sum of the angles in a quadrilateral than the sum of the angles in a right-angled triangle?

A 56°
B 80°
C 90°
D 150°
(E) 180°

The answer is **E**.
Sum of angles in a quadrilateral = 360°
Sum of angles in a right-angled triangle = 180°
360° – 180° = 180°

① What is the value of y in the diagram to the right?

A 40°
B 56°
C 60°
D 44°
E 68°

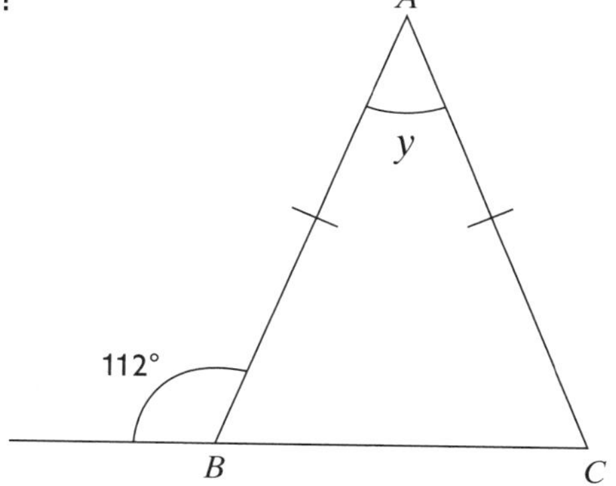

② Which of the following has a different value compared to the others?

A Sum of interior angles of a scalene triangle
B $\frac{\text{Sum of interior angles of a quadrilateral}}{3}$
C An exterior angle of an equilateral triangle × 1.5
D Angles on a straight line
E Sum of interior angles of a triangle

Questions continue on next page

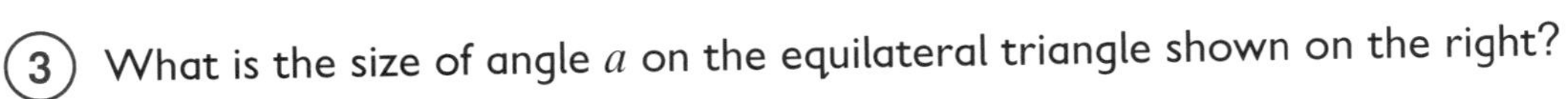

(3) What is the size of angle a on the equilateral triangle shown on the right?

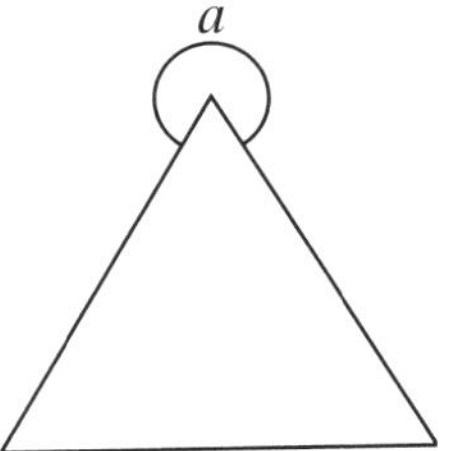

A 140°
B 150°
C 260°
D 300°
E 118°

HELPFUL TIP: *Think about the sum of the internal and external angles around a point.*

(4) What is the size of angle q in the diagram shown on the right?

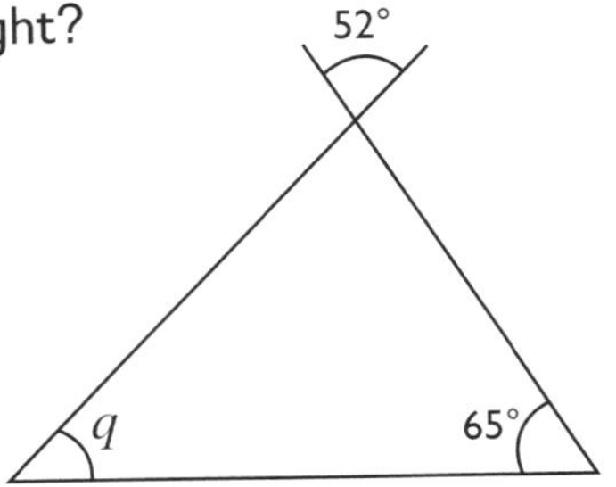

A 63°
B 61°
C 113°
D 117°
E 60°

(5) Figure D consists of a regular pentagon and an isosceles triangle. What is the size of angle k?

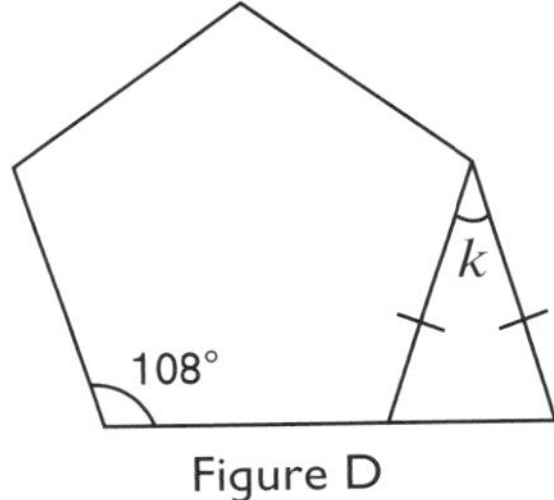

Figure D

A 63°
B 62°
C 20°
D 36°
E 15°

(6) Anne began a journey by walking east for 2 km. She then turned 90° to her right and walked 200 m. Then she again turned 90° to her right and walked 400 m. In which direction is she now facing?

A East
B West
C Southeast
D North
E South

(7) Sachin walks east for 1 km and then turns in a clockwise direction to face southwest. By what angle has he turned?

A 160°
B 180°
C 90°
D 45°
E 135°

(8) What is the reflex angle between the hour and the minute hand at 3:45 p.m.?

A 157.5°
B 202.5°
C 197.5°
D 92.5°
E 257.5°

Use the diagram below to help you answer questions 9 and 10.

The diagram below shows a regular pentagon.

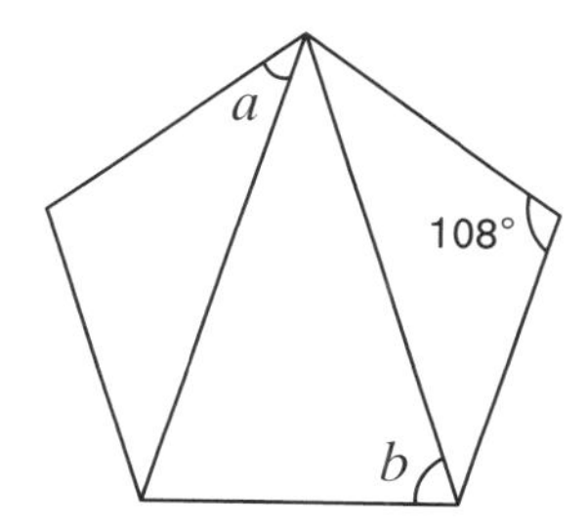

(9) What is the size of angle a?

A 36°
B 72°
C 108°
D 360°
E 45°

(10) Which of the following is $(a + b)$ equal to?

A The interior angle of a square
B The angles on a straight line
C The interior angle of a regular pentagon
D The exterior angle of an equilateral triangle
E A right angle

Test 5: Money

INSTRUCTIONS

You have 10 minutes to complete the following section.
You have 10 questions to complete within the time given.

Write the correct answer in the boxes provided.

EXAMPLE

Rebecca has one of each type of coin in her pocket. What is the total value of the coins in Rebecca's pocket?

The answer is:

£ 3 . 8 8

Sum of all coins = 1 p + 2 p + 5 p + 10 p + 20 p + 50 p + £1 + £2 = **£3.88**

1. A porcelain tile costs £2.45. What is the greatest number of tiles that Max can buy with £300?

2. Kian's car broke down. The repair charges for his car are shown in the table below.

Item	Quantity required	Charge per item
Brake pads	2	£35.75
Wiper blades	2	£22.50
Rear view mirror	1	£13.00
Indicator bulbs	3	£6.10

Kian was also charged for 90 minutes of labour at the rate of £30 per hour. How much did Kian spend in total to repair his car?

£ .

(3) Alastair works for 9 hours over the weekend and 25 hours from Monday to Friday. His rate of pay is shown in the table below.

Weekday	£11.50 per hour
Weekend	£13 per hour

Alastair wishes to buy a laptop that costs £425. How much less are his weekly earnings than the cost of the laptop?

£ ☐☐.☐☐

(4) The cost of 15 litres of petrol is £19. What is the cost of 1 litre of petrol? Round your answer to the nearest 10 p.

£ ☐.☐☐

(5) This is the menu of a cafeteria:

Roll	*75p*
Burger	*£1.20*
Tea	*£1.10*
Coffee	*£1.30*
Sandwich	*£2.30*
Cola	*65p*

Annette buys 3 of an item, 2 of another item and 1 of another item. She pays with a £10 note and receives £4.55 in change. What is the highest priced item that she bought?

☐

(6) An Italian restaurant sells pizzas for £4.75 each. The restaurant sells 100 pizzas per week on average. What is their total profit over 4 weeks if the cost of making 1 pizza is £1.50?

£ ☐,☐☐☐

HELPFUL TIP: *Think about what the relationship is between sales, costs and profit.*

Questions continue on next page

(7) 2 shops, Growell and Greenhouse, sell holly plants as shown below:

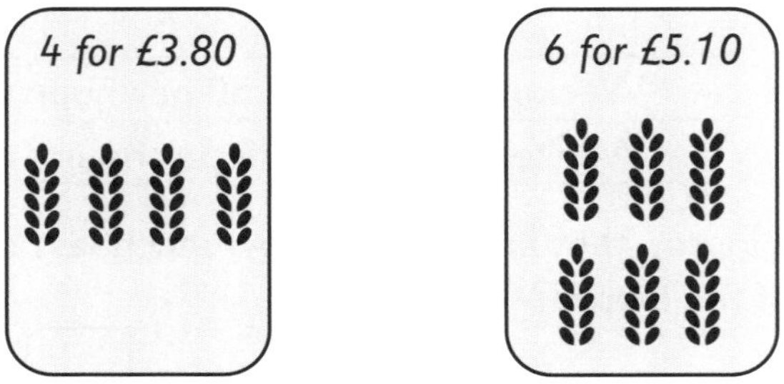

4 for £3.80

Growell

6 for £5.10

Greenhouse

How much money does Mrs Woolley save by buying 24 plants at Greenhouse rather than at Growell?

£ ☐.☐☐

Use the diagram below to help you answer questions 8 and 9.

Imran is exploring the use of a television meter. He can pay for the television he watches in 2 ways, as shown below:

No television meter

£540 per year

With TV meter
£100 per year
+
£0.30 for every hour watched

(8) If Imran watches 3 hours of television per day in 2018, how much will he save annually by installing the television meter?

£ ☐☐☐.☐☐

(9) If Imran watches 5 hours of television per day in 2018, how much money will he spend annually if he installs the television meter?

£ ☐☐☐.☐☐

---

(10) A bookshop sells dictionaries for £3.00 and charges 15% extra for gift-wrapping. If the shop earned £86.25 in total from the sale of dictionaries with gift-wrapping, how many dictionaries did they sell?

☐☐

Test 6: Prime Numbers

INSTRUCTIONS

You have 10 minutes to complete the following section.
You have 10 questions to complete within the time given.

Circle the letter beside the correct answer.

EXAMPLE

How many prime numbers are greater than 8 but less than 18?

A 2
(B) 3
C 4
D 5
E 6

The answer is **B**.
Prime numbers are 11, 13 and 17

(1) What is the answer when the biggest prime number less than 100 is divided by the only even prime number?

A 48.5
B 41.5
C 47.5
D 46
E 49.5

(2) Which of these statements about prime numbers is TRUE?

A The sum of any 2 prime numbers is always even.
B The product of any 2 prime numbers is always odd.
C The product of any 2 prime numbers is always a prime number.
D There are 10 prime numbers less than 30.
E The difference between any 2 prime numbers is always even.

Questions continue on next page

Use the table below to help you answer questions 3 and 4.

This table shows the number of shirts and trousers owned by different boys.

	Shirts	**Trousers**
Aryan	13	21
Rob	19	30
Jake	41	11
Matthew	23	39
Ali	51	33

(3) Who has a prime number of shirts and a prime number of trousers?

A Aryan
B Rob
C Jake
D Matthew
E Ali

(4) What is the difference between the number of shirts and trousers owned by the boy who has a composite number of each?

A 8
B 11
C 30
D 18
E 16

---

(5) This table shows 3 numbers. Ralph picks out the prime numbers and adds them together. What is Ralph's answer?

179	289	347

A 526
B 815
C 412
D 310
E 302

(6) In writing the prime factors of 30, Jacob missed out 1 number. He has written 2 and 5. What number has he missed?

A 3
B 4
C 9
D 7
E 11

(7) This table lists some numbers.

350	480	640	1,290	3,090

Which of the following explains why none of these numbers is prime?

A They are all even numbers and no even number is prime.
B 10 is a factor of each of these numbers so they cannot be prime.
C They are all greater than 100 so they cannot be prime.
D They are all divisible by 4 so they cannot be prime.
E The sum of their digits is even so they cannot be prime.

(8) Square numbers can be written as a product of prime factors. Example: $64 = 2^6$
How can 36 be expressed in the same way?

A $2^3 \times 2^2$
B $2^2 \times 3^2$
C $9^2 \times 3^2$
D $2^3 \times 3^2$
E $2^3 \times 3^3$

(9) 12 chairs are arranged in rows of equal length. How many possible arrangements will result in a prime number of chairs in each row?

A 8
B 6
C 3
D 4
E 2

(10) Prime numbers that are either 2 greater or 2 smaller than another prime number are called 'twin prime numbers'. Which of these is not a twin prime number pair?

A 59 and 61
B 11 and 13
C 109 and 111
D 41 and 43
E 137 and 139

Test 7: Percentage

INSTRUCTIONS

You have 10 minutes to complete the following section.
You have 10 questions to complete within the time given.

Circle the letter beside the correct answer.

EXAMPLE

There are 50 birds in a cage. 20% of these birds are red.

How many red birds are there in the cage?

A 20
B 30
(C) 10
D 50
E 40

The answer is **C**.
Number of red birds = 20% of 50 = 10

(1) Sanya bought a pair of trainers for £34, which was a 15% reduction from the original price. What was the original price of the trainers?

A £42
B £51
C £40
D £56
E £64

(2) This table shows the favourite cereals of a group of students.

If 15% of the students chose Bran Flakes, how many students chose Honey Loops?

Cinnamon Puffs	12
Bran Flakes	9
Oats	15
Choco Pops	11
Honey Loops	?

A 13
B 15
C 11
D 10
E 16

3. Wendy's wedding dress cost £320 plus 17% tax. Her wedding ring cost £290 plus 18% tax. How much more did she pay in total for her dress than for her wedding ring?

A £32
B £31.20
C £32.30
D £30.80
E £32.20

4. On Amy's bookshelf, 20% of the 220 books are non-fiction. $\frac{5}{11}$ of these non-fiction books are biographies. How many biographies are there on Amy's bookshelf?

A 15
B 12
C 24
D 13
E 20

5.

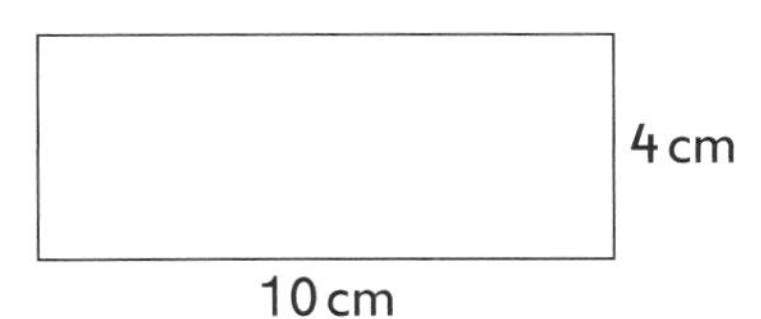

If the sides of the above rectangle are each enlarged by 20%, what is the area of the enlarged rectangle?

A 57.6 cm^2
B 27.6 cm^2
C 58.6 cm^2
D 36.6 cm^2
E 57.2 cm^2

6. $30\% \times a = b$

Which of these statements is TRUE?

A $\frac{a}{b} = 0.3$
B $\frac{a}{b} = 0.7$
C $a = \frac{10b}{3}$
D $b = \frac{a}{0.3}$
E None of the above

Questions continue on next page

(7) The price of a hotel room in June is £x. The price is decreased by 20% in July and then increased by 25% in August. What is the percentage change in price from June to August?

A There is no percentage change in price
B The price increases by 5%
C The price decreases by 5%
D The price increases by 15%
E The price increases by 45%

Use the pie chart below to help you answer questions 8 and 9.

This pie chart shows the nationalities of the population of a certain island.

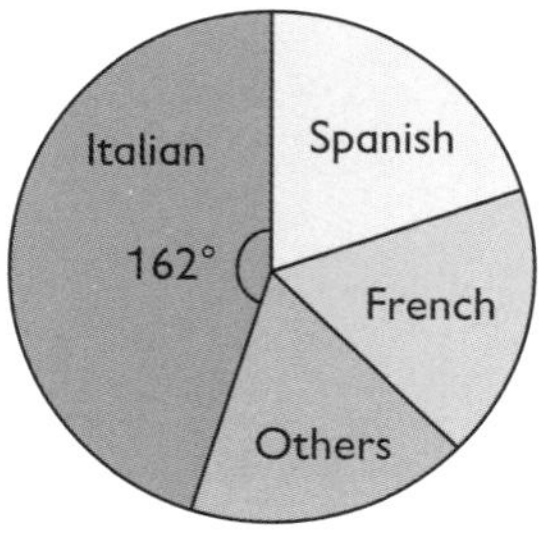

(8) What percentage of the population is Italian?

A 35%
B 12%
C 50%
D 56%
E 45%

(9) Apart from the Italian segment, the other segments on the pie chart are all equal in size. If the island has a population of 7,500, how many French people live on the island?

A 1,650
B 2,250
C 1,220
D 1,375
E 1,350

---

(10) Frank's income is $\frac{5}{10}$ as great as Alfred's. By what percentage must Frank's income increase so that it is equal to Alfred's?

A 50%
B 40%
C 100%
D 75%
E 200%

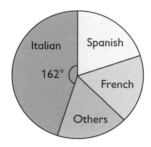

Test 8: Probability

INSTRUCTIONS

You have 10 minutes to complete the following section.
You have 10 questions to complete within the time given.

Write the correct answer in the boxes provided (one digit per box).

EXAMPLE

What is the probability of rolling a fair dice and scoring a 4?

The answer is: $\frac{1}{6}$

Rolling a 4 is 1 out of 6 equally possible outcomes, so probability is $\frac{1}{6}$

1. Anna calculates the probability of a month having more than 28 days in the year 2020. What is the answer to her calculation?

HELPFUL TIP: *Think about the name given to years in which all months have more than 28 days.*

2. Jack sees the following message on his computer:

OPEN YOUR WEB BROWSER

What is the probability that 1 letter selected from this message at random is a consonant?

Questions continue on next page

3. 10% of people are left-handed. Out of these people, a fifth are ambidextrous, which means they are comfortable using either their left or right hand. What is the probability that a person is ambidextrous?

☐.☐☐

4. Sharon selects a shape at random from the ones below:

What is the probability that the shape she selects has no lines of symmetry?

☐.☐

5. Danny has green, yellow and blue ducks. He has twice as many blue ducks as green ducks. If the probability of choosing a yellow duck is $\frac{5}{14}$, what is the smallest number of green ducks he could have?

☐

6. Philip polled Year 3, 4 and 5 students to find out their favourite after-school activity. The results are shown below:

	Swimming	**Drama**	**Gymnastics**
Year 3	12	3	4
Year 4	8	12	7
Year 5	1	10	3

What is the probability, rounded to 2 decimal places, that the favourite activity of a student selected at random is Drama?

☐.☐☐

7. The probability that a resident of Berlin likes summer is $\frac{4}{5}$. The probability that a resident of Berlin likes thunderstorms is $\frac{3}{10}$. What is the probability that a resident of Berlin likes summer and thunderstorms?

☐.☐☐

(8) The diagram below shows a fair 8-sided spinner.

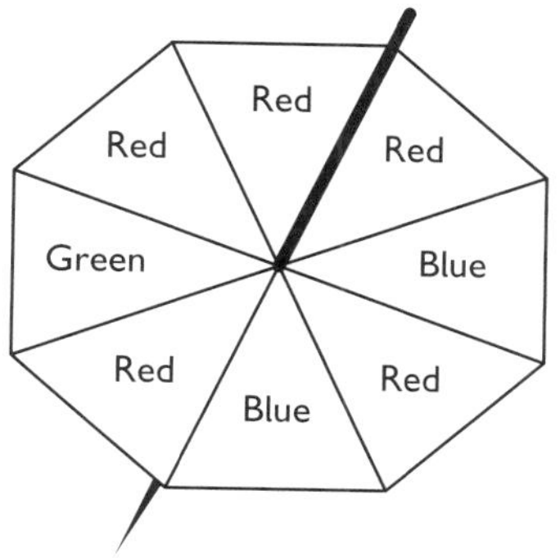

Red | Red | Red | Blue | Red | Blue | Red | Green

Esther paints the green section of the spinner red. What is the probability of landing on a red section of the newly painted spinner?

☐.☐☐

(9) This table shows the hair colour of 10 students in a class. If one student is picked at random, what is the probability that they have black hair?

Student	Hair colour
1	black
2	blonde
3	ginger
4	brown
5	brown
6	black
7	blonde
8	ginger
9	ginger
10	blonde

☐.☐

(10) The numbers below are the actual weights of 6 bags of rice that are each supposed to weigh 5 kg.

5.15 kg   5.02 kg   4.98 kg   5.200 kg   6.0 kg   4.78 kg

What is the probability that a bag selected at random contains more than 200 g less rice than it is supposed to?

$\frac{☐}{☐}$

Test 9: Ratio

INSTRUCTIONS

You have 10 minutes to complete the following section.
You have 10 questions to complete within the time given.

Circle the letter beside the correct answer.

EXAMPLE

The ratio of boys to girls in a class is 2:1. If there are 8 girls in the class, how many boys are there in the class?

(A) 16
B 8
C 24
D 12
E 10

The answer is **A**.
boys:girls = 2:1
So number of boys = number of girls × 2
8 × 2 = 16

1. Simon and Chloe divide £30 in the ratio 5:1. Simon uses 20% of his share to buy a book. How much money does he have left?

A £12
B £25
C £6
D £20
E £3

2. The perimeter of this rectangle is 96 cm. If the width and length are in the ratio 1:2, what is the length of the rectangle?

A 14 cm
B 32 cm
C 2 cm
D 7 cm
E 1 cm

3. For every 6 rooms in an office, 2 will have carpeted floors. If there are 360 rooms in the office in total and each carpeted room requires 20 m^2 of carpet, what is the total area of carpet required for the office rooms?

A 2,000 m^2
B 120 m^2
C 2,400 m^2
D 12 m^2
E 240 m^2

4. Ellen takes the piece of string shown below. It has a length of $3y$ cm.

She cuts 12 cm from it and realises that the piece she has cut is shorter than the piece that remains. What is the ratio of the length of the longer piece to the shorter piece?

A $(3y - 12):12$
B 1:4
C $12:(3y - 12)$
D $(3y + 12):12$
E $12:3y$

5. Owen can answer 35 questions in 12 minutes. Working at this speed, how long does he need to answer all 140 questions in a test?

A 2 hours
B 48 minutes
C 2.1 hours
D 0.6 hours
E 84 minutes

6. A company plans to hire 4 people to complete a task in 15 days, each working for 8 hours a day. The company decides to hire 12 people instead, each working 10 hours a day. In how many days will the task now be completed?

A 6 days
B 16 days
C 15 days
D 10 days
E 4 days

Questions continue on next page

Use the diagram below to help you answer questions 7 and 8.

Christopher's new bathroom contains a pattern of square grey and white tiles on slabs. This diagram shows the design on 1 slab:

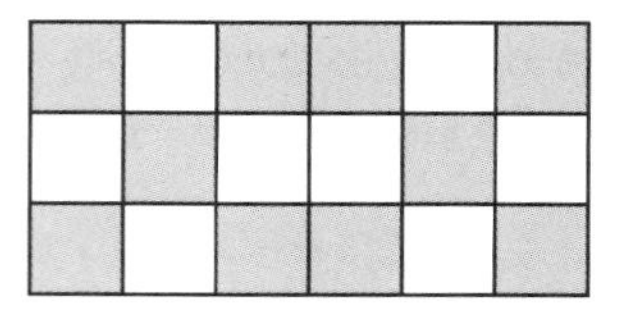

7. If the side of 1 square tile measures 20 cm and he uses 20 slabs, what is the total area of all the grey tiles used?

A 8 m^2
B 800 cm^2
C 800,000 cm^2
D 0.8 m^2
E 80 m^2

8. If the cost of a grey tile is £2.10 and the cost of a white tile is £0.50, how much does he spend altogether?

A £100
B £320
C £500
D £650
E £210

Use the recipe below to help you answer questions 9 and 10.

This recipe will make 6 pies:

For 6 pies
240 g butter
300 g sugar
210 g flour
183 g meat

9. What is the ratio of butter to meat in the recipe?

A 4:9
B 183:240
C 40:30
D 8:30
E 80:61

10. Emma wants to make 19 pies. Flour is sold in 250 g packets. How many packets of flour does she need?

A 3
B 5
C 2
D 10
E 6

Test 10: Perimeter

INSTRUCTIONS

You have 10 minutes to complete the following section.
You have 10 questions to complete within the time given.

Circle the letter beside the correct answer.

EXAMPLE

A square has a side length of 14 cm.

What is the perimeter of the square?

A 16 cm
B 14 cm
C 28 cm
(D) 56 cm
E 80 cm

The answer is **D**.
Side length × 4 = perimeter
14 cm × 4 = 56 cm

(1) Sid jogs 6 times around a rectangular field with a length of 1.5 km. If he jogs 24 km in total, what is the width of the field?

A 2 km
B 1 km
C 500 m
D 250 m
E 3.5 km

(2) 2 pieces of cloth are shown below. Cloth A is a rectangle and Cloth B is a square.

Which cloth has a greater area and by how much?

A Cloth A by 41 cm^2
B Cloth A by 4,100 cm^2
C Cloth A by 0.41 cm^2
D Cloth B by 4,100 cm^2
E Cloth B by 100 cm^2

1 m | Cloth A | 4 m

Cloth B | 210 cm

Questions continue on next page

3. A square sheep pen has a side length of 220 m. What is the cost of placing a fence all around the sheep pen if the cost of fencing is £65 per kilometre?

A £58.20
B £57,200
C £52.10
D £64,800
E £57.20

4. The shape below has 2 lines of symmetry and a perimeter of 2.15 km.

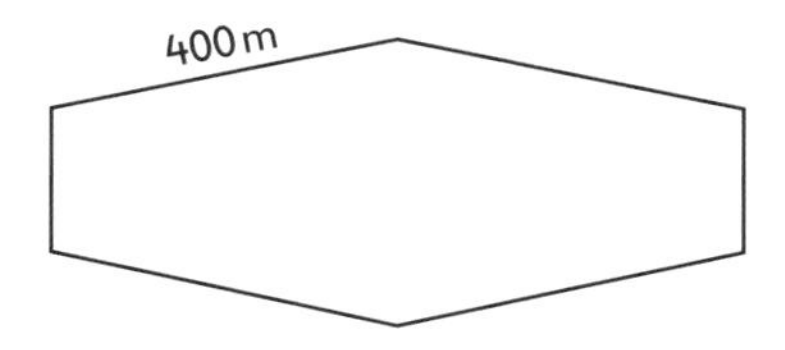

What is the length of the shortest side of the shape?

A 550 m
B 275 m
C 0.55 km
D 2.75 km
E 215 m

5. Which of these statements is TRUE?

A If the side length of a square doubles, its perimeter triples
B If the side length of a square doubles, its perimeter doubles
C If the side length of a square doubles, its perimeter quadruples
D If the side length of a square halves, its perimeter doubles
E If the side length of a square doubles, its perimeter decreases by one third

6. What is the ratio in its simplest form of the perimeters of the regular hexagon and the equilateral triangle below if they have the same side length?

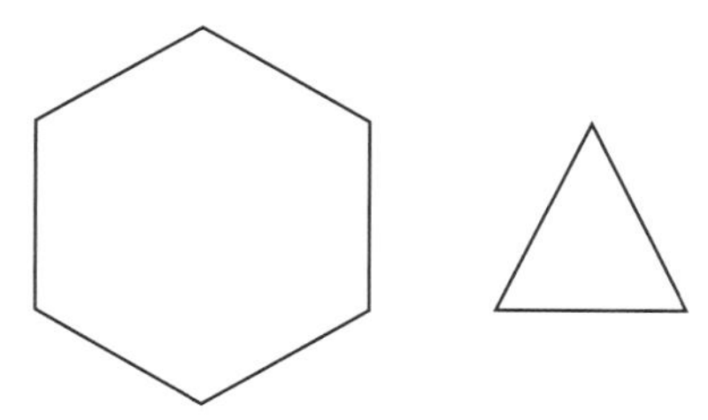

A 1:3
B 6:4
C 2:1
D 4:6
E 3:1

7. Which of these triangles has the greatest perimeter?

- **A** A scalene triangle with sides measuring 224 mm, 108 mm and 150 mm
- **B** An equilateral triangle with a side length of 1.85 cm
- **C** An isosceles triangle with sides measuring 155 mm, 155 mm and 30 mm
- **D** A right-angled triangle with sides measuring 0.75 cm, 1 cm and 1.25 cm
- **E** A triangle in which the smallest side measures 0.5 cm, the largest side measures 1.5 cm and the other side is twice as large as the smallest side

8. The diagram below shows a grey square placed on top of a white square.

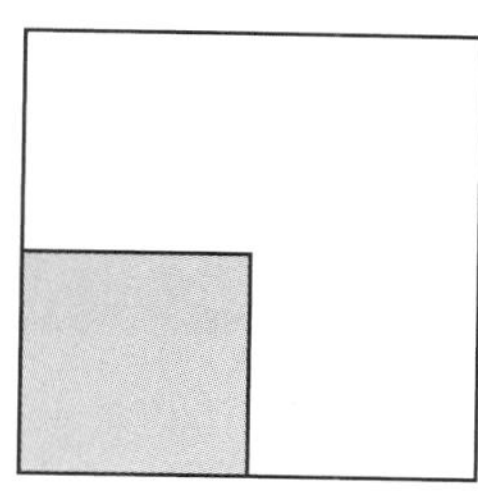

The area of the white square that is not covered by the grey square is 675 cm^2.

Which of the following shows the perimeters of the 2 squares?

- **A** Grey square: 8 cm; White square: 16 cm
- **B** Grey square: 60 cm; White square: 120 cm
- **C** Grey square: 12 cm; White square: 56 cm
- **D** Grey square: 40 cm; White square: 24 cm
- **E** Grey square: 32 cm; White square: 20 cm

9. If x represents the width of Square A and y represents the length of Square A, which of the following expressions is TRUE?

- **A** $xy = x + y$
- **B** $2xy = x + y$
- **C** $\frac{xy}{3} = x + y$
- **D** $xy = x + \frac{y}{2}$
- **E** $4y = x + y + 2x$

10. Robert swims 4 widths of the pool shown below. He then swims twice this distance in lengths. How many lengths of the pool does Robert swim?

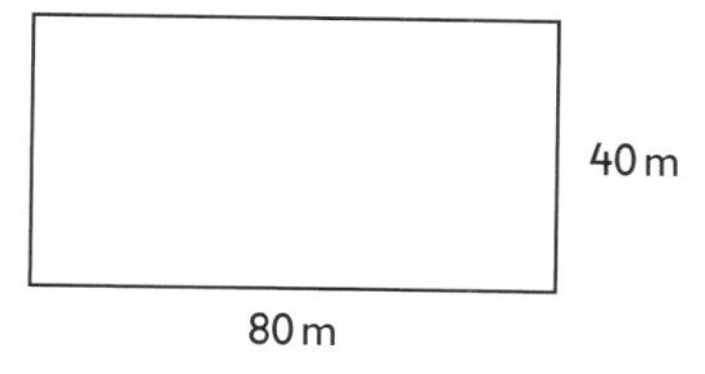

- **A** 1
- **B** 2
- **C** 4
- **D** 8
- **E** 16

Test 11: Graphs and Charts

INSTRUCTIONS

You have 9 minutes to complete the following section.
You have 9 questions to complete within the time given.

Circle the letter beside the correct answer.

EXAMPLE

This timetable shows the schedule of 2 trains.

How much longer does it take Train B than Train A to travel from Frochester to Gochester?

	Train A	Train B
Frochester	17:34	19:21
Bochester	17:49	19:40
Gochester	18:03	19:59
Lochester	18:11	20:08

A 7 minutes
B 8 minutes
(C) 9 minutes
D 10 minutes
E 11 minutes

The answer is **C**.
Train A: 17:34 → 18:03 is 29 minutes
Train B: 19:21 → 19:59 is 38 minutes
38 minutes – 29 minutes = 9 minutes

(1) This pie chart shows the results of a survey asking people why they went on holiday.

How many different responses are represented by an acute-angled segment?

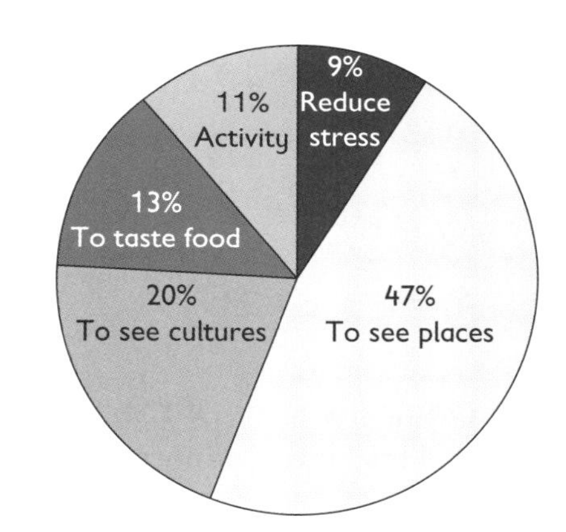

A 1
B 2
C 3
D 4
E None

(2) This table shows the areas of the gardens of different houses along a street.

If the range of the values is 32 m^2 and House 1 has the biggest garden, what is the area of the garden of House 6?

House 1	117 m^2
House 2	111 m^2
House 3	90 m^2
House 4	110 m^2
House 5	100 m^2
House 6	?

A 17 m^2
B 30 m^2
C 85 m^2
D 68 m^2
E 55 m^2

Use the table below to help you answer questions 3 and 4.

This table shows the percentage of sales per category for an online store.

Department	% of total sales	
	2016	**2017**
Food	32%	18%
Electrical	12%	26%
Clothing	35%	38%
Leisure	21%	18%

3. Connor wants to display this information visually. Which of the following is the most appropriate way of representing the information?

A A histogram
B A tally chart
C A frequency diagram
D A double bar graph
E A scatter diagram

4. Connor draws a pie chart to display the sales figures for 2016. What is the size of the angle that represents Clothing?

A 108°
B 144°
C 126°
D 90°
E 45°

Use the graph below to help you answer questions 5 and 6.

This graph shows the number of adult and child patients that were treated at a First Aid unit over a 10-day period.

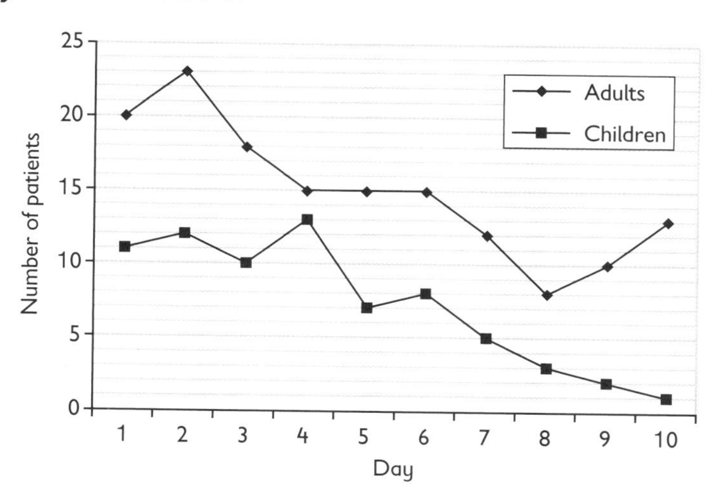

5. What is the range of the number of child patients per day?

A 12
B 13
C 1
D 14
E 5

Questions continue on next page

6. What is the smallest difference between the number of adult and child patients treated on any single day?

A 1
B 2
C 4
D 8
E 6

Use the notice below to help you answer questions 7 and 8.

This notice shows the opening times of a customer service centre.

Customer service opening times

- Monday to Friday 8 a.m. – 8 p.m.
- Saturday 8 a.m. – 5 p.m.
- Sunday 10 a.m. – 4 p.m.

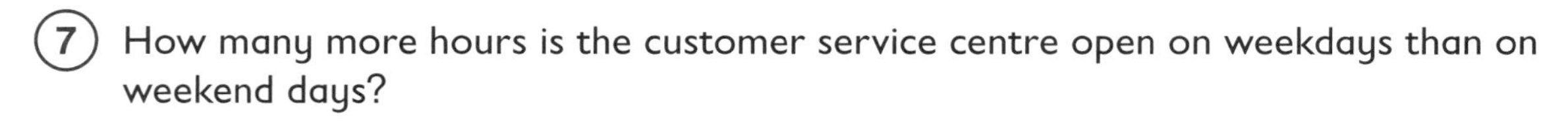

7. How many more hours is the customer service centre open on weekdays than on weekend days?

A 20
B 25
C 60
D 45
E 12

8. Ramona works part-time at the customer service centre. She works all available hours on Sundays and Mondays. If she gets paid £11 per hour at the weekend and £9.20 per hour on weekdays, how much does she earn in a fortnight?

Round your answer to the nearest 100 pounds.

A £400
B £300
C £350
D £200
E £600

9. This graph shows a relationship between distance and time.

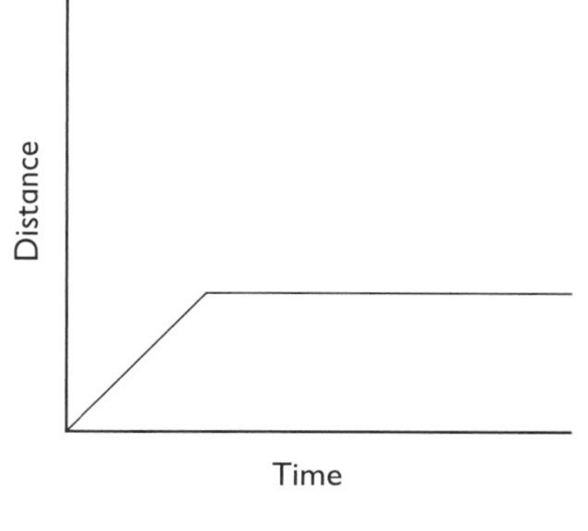

Which of the following could this graph possibly represent?

A A child running at a constant speed
B A child coming to rest after running at a variable speed
C A child running at a constant speed and then stopping
D A child at rest
E A standing child

Test 12: Tables

INSTRUCTIONS

You have 10 minutes to complete the following section.
You have 10 questions to complete within the time given.

Write the correct answer in the boxes provided (one digit per box).

EXAMPLE

What percentage of these shapes are black?

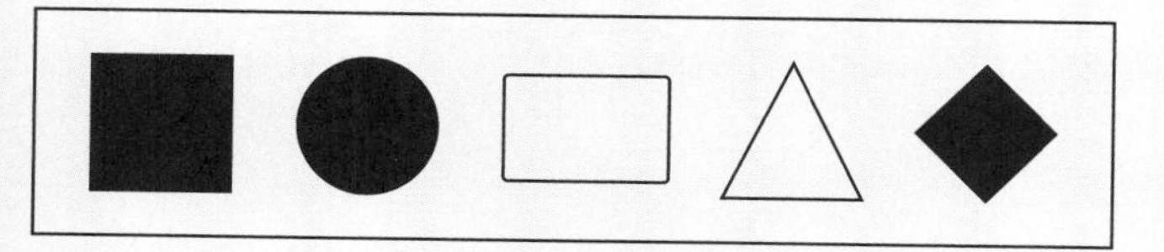

The answer is:

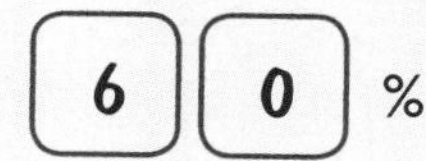

3 out of 5 shapes are black

$\frac{3}{5} \times 100 = \mathbf{60\%}$

Use the diagram below to help you answer questions 1 and 2.

This diagram contains 9 shapes.

① 1 shape is selected at random. What is the probability that it has exactly 4 lines of symmetry?

② Which number should replace Y in the table below?

White shapes with no lines of symmetry	Shapes with more than two lines of symmetry	Black shapes with no lines of symmetry	Shapes with no straight lines
1	3	Y	2

Questions continue on next page

Use the timetable below to help you answer questions 3 and 4.

Andrew is planning a trip to the York Racecourse Model Railway Exhibition.

Buses to York Racecourse Model Railway Exhibition

York Station Road	0938	0953	1008	1023	1038	1053	**then every 15 mins at**	08	23	38	53
York Railway Station	0945	1000	1015	1030	1045	1100		15	30	45	00
York Racecourse	0953	1008	1023	1038	1053	1108		23	38	53	08

Buses from York Racecourse Model Railway Exhibition

York Racecourse	0955	1010	1025	1040	1055	1053	**then every 15 mins at**	10	25	40	55
York Railway Station	1005	1020	1035	1050	1105	1100		20	35	50	05
York Station Road	1007	1022	1037	1052	1107	1108		22	37	52	07

(3) The exhibition is a 12-minute walk from York Racecourse. If Andrew wants to attend the 10:15 show at the exhibition, at what time should he board the bus at York Station Road?

☐☐ : ☐☐

(4) Andrew decides to meet his friend for dinner at 7:15 p.m. at York Station Road. What is the latest bus he can board from York Racecourse and still be on time?

☐ : ☐☐ p.m.

Use the table below to help you answer questions 5 and 6.

This table shows the task plan for children at a holiday camp, starting at 10:00 a.m.

Task	Duration
Morning warm-up	15 minutes
News roundup	22 minutes
Team-building activity	2.4 hours
Lunch break	40 minutes
Sports roundup	1.8 hours
Windup	1 minute

(5) At what time does lunch break end?

☐☐ : ☐☐

(6) How long does the sports roundup last?

☐ hour ☐☐ minutes

(7) This table shows the hours worked by a teacher on 3 days. There was a 1-hour gap for lunch break each day.

Day	Start Time	Finish Time
Monday	08:10	15:35
Tuesday	08:30	15:20
Wednesday	07:45	15:30

Excluding lunch breaks, how many hours did the teacher work in total over the 3 days?

☐☐ hours

Use the table below to help you answer questions 8 and 9.

Highest Common Factor of 900 and 250	a
Lowest Common Multiple of 25 and 60	b
Lowest Common Multiple of x and y	56

(8) What is $\frac{b}{a}$?

☐

(9) Which of the following are possible values of x and y?

i. $x = 8$, $y = 12$

ii. $x = 8$, $y = 14$

iii. $x = 10$, $y = 12$

iv. $x = 8$, $y = 18$

v. $x = 7$, $y = 6$

$x =$ ☐ $y =$ ☐

---

(10) This table shows the time 3 buses depart from Gatwick and when they arrive at Luton.

Bus	A	B	C
Gatwick	06:35	07:00	07:15
Luton	08:09	08:39	?

If the average time taken for the journey by the 3 buses is 1 hour 52 minutes, at what time does Bus C reach Luton?

☐☐ : ☐☐

Test 13: Time

INSTRUCTIONS

You have 8 minutes to complete the following section.
You have 8 questions to complete within the time given.

Circle the letter beside the correct answer.

EXAMPLE

A puzzle takes Tom 3 hours and 40 minutes to complete.

How many minutes does it take Tom to complete the puzzle?

A 120 minutes
B 160 minutes
C 180 minutes
(D) 220 minutes
E 240 minutes

The answer is **D**.

1 hour = 60 mins

3 hours 40 mins = (3 × 60) mins + 40 mins = 220 mins

(1) A Year 2 class has 35 minutes of phonics lessons every day. The autumn term has 8 weeks. How long do they spend on phonics in the autumn term assuming that each week has 5 schooldays? Round your answer to the nearest hour.

A 21 hours
B 24 hours
C 23 hours
D 70 hours
E 30 hours

Use the diagram below to help you answer questions 2 and 3.

These clocks show the simultaneous time in London and Beijing.

London (p.m.)

Beijing (p.m.)

2) Lily lives in London and her grandad lives in Beijing. Lily calls her grandad at 3:10 p.m. London time. Her grandad's clock shows 21:45. Is her grandad's clock fast or slow and by how many minutes?

- **A** Slow by 35 minutes
- **B** Slow by 25 minutes
- **C** Fast by 35 minutes
- **D** Fast by 25 minutes
- **E** Slow by 30 minutes

3) Lily's grandad watches a film that lasts for 2 hours and 15 minutes. If the film starts at 10:20 p.m. Beijing time, what time will it be in London when the film ends?

- **A** 6:35 a.m.
- **B** 7:35 a.m.
- **C** 16:35
- **D** 18:35
- **E** 17:35

Use the table below to help you answer questions 4 and 5.

This table shows the time taken by different students to complete a 500 m running race.

Student	Time
Stefan	5.2 minutes
Sia	5 minutes 6 seconds
Hannah	0.1 hours
Jenny	330 seconds
Meera	512 seconds

4) Who won the race?

- **A** Stefan
- **B** Sia
- **C** Hannah
- **D** Jenny
- **E** Meera

5) What is the mean time taken by the students to complete the race?

- **A** 360 seconds
- **B** 366 seconds
- **C** 368 seconds
- **D** 362 seconds
- **E** 364 seconds

Questions continue on next page

Use the timetable below to help you answer questions 6, 7 and 8.

This is the Year 6 timetable at Bell School.

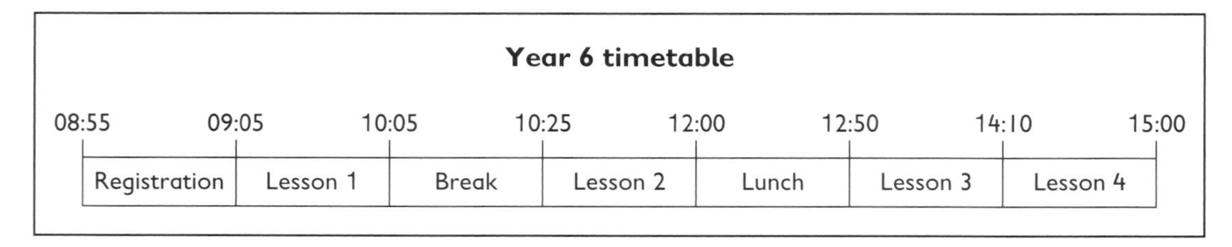

At Bell School, Year 1 starts at the same time as Year 6, and has the same lesson duration. However, Year 1 Breaks are 10 minutes longer and so is Lunch.

(6) At what time does Lesson 2 end for Year 1?

A 12:05 p.m.
B 12:15 p.m.
C 12:10 a.m.
D 12:10 p.m.
E 10:35 a.m.

(7) At what time does Lesson 4 start for Year 1?

A 2:30 p.m.
B 2:20 p.m.
C 2:10 a.m.
D 3:10 p.m.
E 2:35 p.m.

(8) Kyra is in Year 6 while her brother is in Year 1. How many minutes earlier does she finish the schoolday compared to her brother?

A 10 minutes
B 15 minutes
C 40 minutes
D 20 minutes
E 5 minutes

Test 14: Weights and Measures

INSTRUCTIONS

You have 10 minutes to complete the following section.
You have 10 questions to complete within the time given.

Circle the letter beside the correct answer.

EXAMPLE

How many grams are there in 3.52 kg?

A 35.2 g
(B) 3,520 g
C 352 g
D 352.2 g
E 35,200 g

The answer is **B**.
1 kg = 1,000 g
3.52 kg × 1,000 = 3,520 g

(1) Tasnin adds together all the values shown in this table.

11 mm	14 cm	2.1 m	0.002 km	9.5 cm

What is her answer in mm?

A 4.346 mm
B 43.46 mm
C 123.8 mm
D 4,346 mm
E 2,098 mm

(2) A jug with a capacity of 3 litres is 40% filled with juice. How many 50 ml cups can be poured from the jug of juice?

A 16 cups
B 24 cups
C 28 cups
D 20 cups
E 15 cups

Questions continue on next page

Use the diagram below to help you answer questions 3 and 4.

This table shows the properties of a £2 coin.

Diameter	28.4 mm
Weight	12.0 g
Thickness	2.50 mm

(3) Lauren has a number of £2 coins. Together they weigh 348 g. How many more coins does she need to make a total of £70?

A 29
B 11
C 6
D 4
E 12

(4) If Lauren places £70 worth of £2 coins end to end in a straight line as shown below, what is the total length of the line in cm?

A 102.24 cm
B 99.4 cm
C 11 cm
D 199.9 cm
E 98.8 cm

---

(5) A map is drawn to a scale of 1:1,200. What distance on the map in cm would represent an actual distance of 1.44 km?

A 120 cm
B 100 cm
C 12 cm
D 10 cm
E 180 cm

(6) What is the value of 0.08 tonnes in grams?

A 8,000 grams
B 5,000 grams
C 80,000 grams
D 4,000 grams
E 800 grams

7. The diagram below shows a square being measured by a ruler.

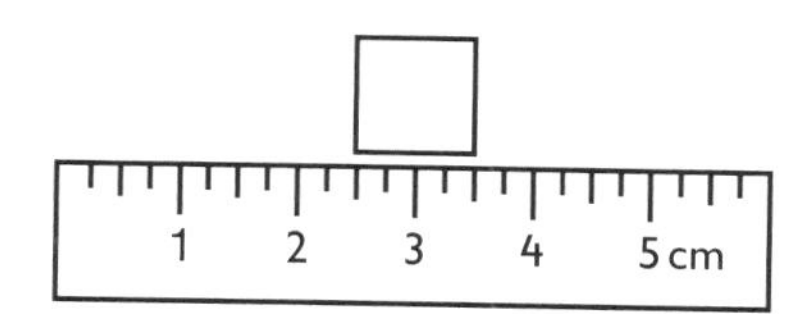

Which of the following statements is TRUE?

A The perimeter of the square is 1 cm
B The perimeter of the square is twice its width
C The area of the square is 1 cm^2
D The perimeter of the square is 3.2 cm
E None of the above

8. Yvonne bought some cinnamon. The weight of the cinnamon is shown on the scale below. If cinnamon costs £25 per kg, how much did she pay?

Round your answer to the nearest 10 p.

A £1.60
B £3.00
C £1.80
D £2.20
E £9.90

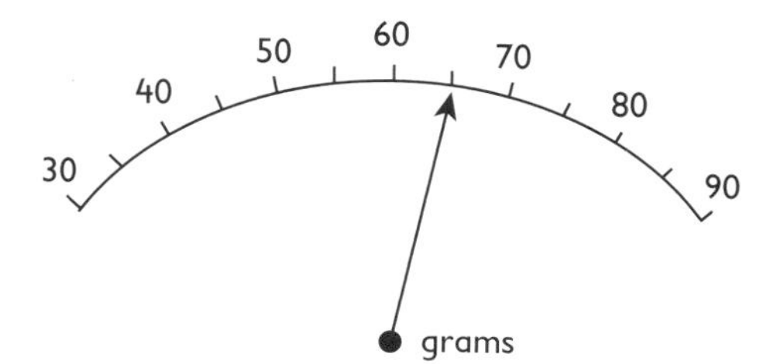

9. Kieran's luggage weighs a whole number of grams. Its weight when rounded to the nearest kilogram is 29 kg. What is the difference between the greatest possible weight and the smallest possible weight of his suitcase?

A 900 g
B 950 g
C 800 g
D 999 g
E 500 g

10. The scale below shows a monkey's weight in kilograms 5 months ago. The monkey became ill and its weight decreased by 20% over the next 5 months. The monkey lost an equal amount of weight each month. How much weight did the monkey lose per month?

A 180 g
B 1,750 g
C 1.76 kg
D 2.2 kg
E 1.076 kg

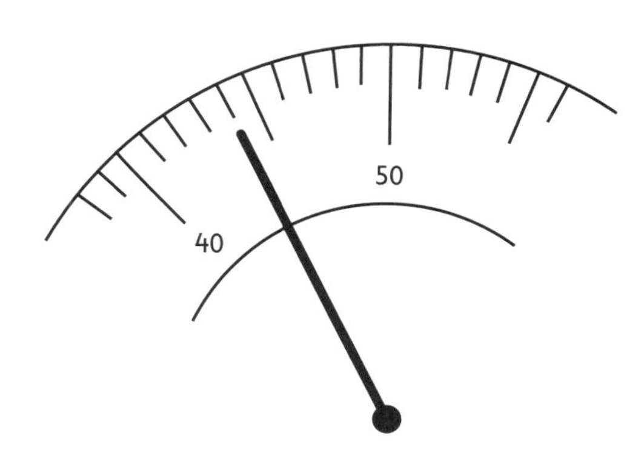

Test 15: Area

INSTRUCTIONS

You have 10 minutes to complete the following section.
You have 10 questions to complete within the time given.

Write the correct answer in the boxes provided (one digit per box).

EXAMPLE

The length of a rectangle is double its width. If the rectangle has a width of 7 cm, what is the area of the rectangle?

The answer is:

9 8 cm^2

Length = 7 cm × 2 = 14 cm
Area = length × width = 14 cm × 7 cm = **98 cm^2**

1. The base of a triangle is 14.2 cm. If the area of the triangle is 106.5 cm^2, what is the height of the triangle?

   ☐☐ cm

Use the diagram below to help you answer questions 2 and 3.

This diagram shows a rectangular park that contains a grey concrete path.

200 m

45 m

2. What is the width of the concrete path if its area is 112.5 m^2?

   ☐.☐ m

3. Apart from the concrete path, the rest of the park is covered by grass. Approximately what percentage of the park is covered by grass?

   ☐☐ %

(4) The length of a rectangle is halved and its width is tripled. What is the percentage increase in the rectangle's area?

☐☐ %

(5) The area of the isosceles triangle below is 75 cm^2. What is the perimeter of the triangle?

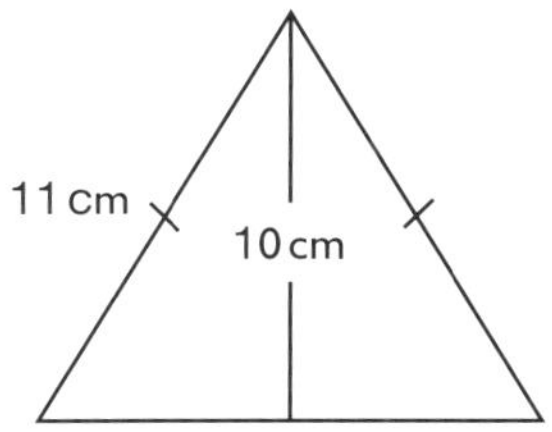

11 cm

10 cm

☐☐ cm

(6) This diagram shows an H-shape drawn on a rectangular grid consisting of rectangles with a length of 2 cm and a width of 1.5 cm.

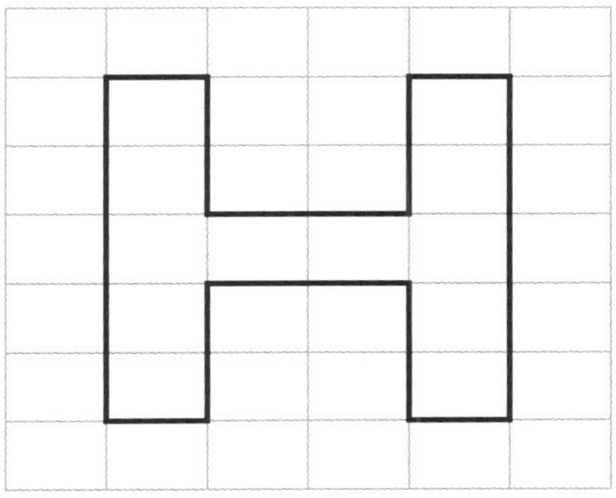

What is the area of the H-shape?

☐☐ cm^2

(7) The cost of building a fence around the square lawn shown below is £120 when using fencing that costs £6 per metre of perimeter. What would the cost of building the fence be if the fencing used cost £9.50 per m^2 of area?

LAWN

£ ☐☐☐.☐☐

Questions continue on next page

Use the diagram below to help you answer questions 8 and 9.

This diagram shows the square outer frame of a rectangular picture.

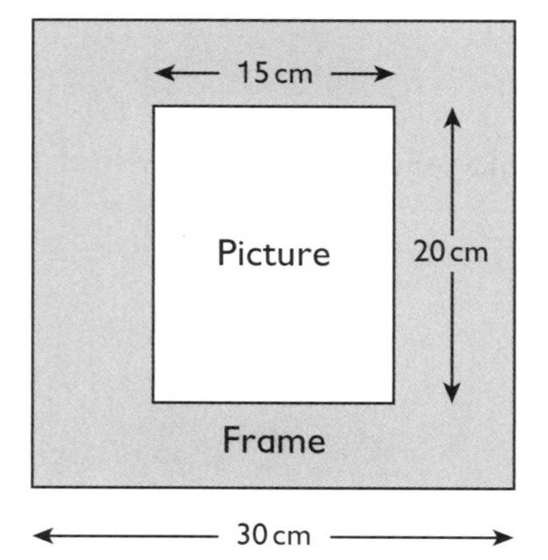

(8) Adriana decorates the frame with 4 buttons for every 10 cm² of the frame.
How many buttons does she use in total to decorate the frame?

(9) Adriana wishes to enlarge the picture and the frame by a scale factor of 4.
What would the enlarged area of the picture be?

---

(10) The perimeters of 2 squares are in the ratio 10:4. What is the ratio of their areas?

HELPFUL TIP: *Remember to take note of the order of the values in a ratio.*

Test 16: Volume

INSTRUCTIONS

You have 10 minutes to complete the following section.
You have 10 questions to complete within the time given.

Write the correct answer in the boxes provided (one digit per box).

EXAMPLE

What is the volume of a cube with a side length of 3 cm?

The answer is:

cm^3

Volume = length × width × height = 3 cm × 3 cm × 3 cm = **27 cm^3**

1. The dimensions of the matchbox shown below are 8 cm × 5 cm × 3 cm.

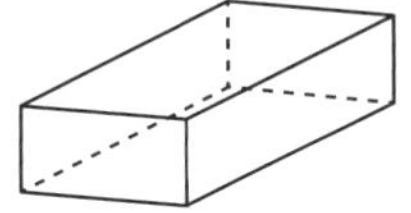

What is the total volume of 20 identical matchboxes?

cm^3

2. A machine for digging costs £12.50/m^3 of earth removed. What is the cost of using the digging machine to remove earth to form the cube-shaped pit shown below?

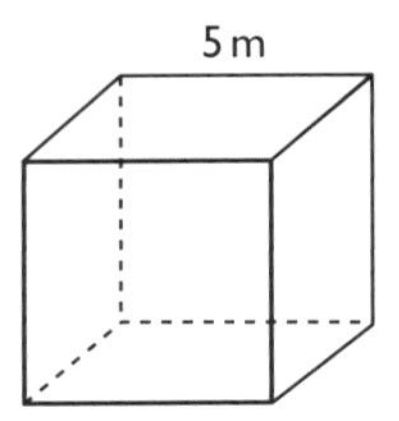

£ 

Questions continue on next page

Use the diagram below to help you answer questions 3 and 4.

The staircase below is built from identical cubes.

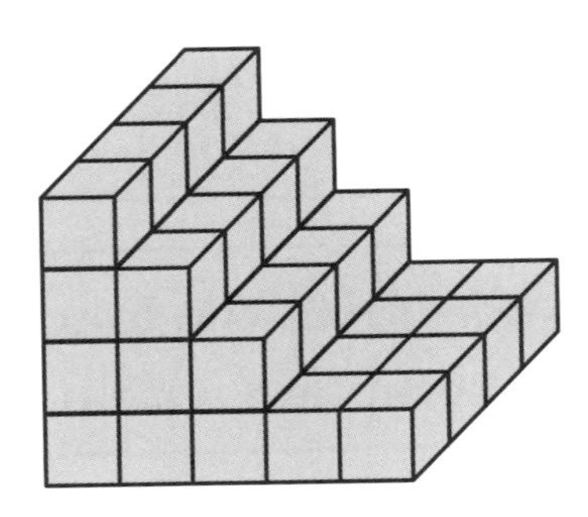

(3) If the total volume of the staircase is 9,504 cm^3, what is the side length of each cube?

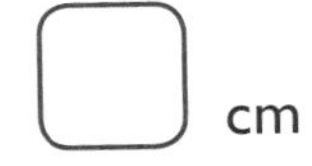

(4) What fraction of the volume of the staircase is made up by the cubes in the lowest layer?

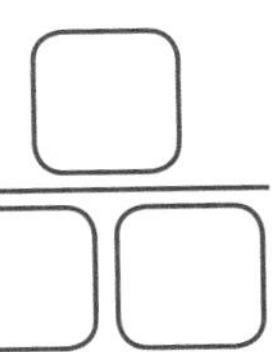

---

(5) A warehouse in the form of a cuboid measures 60 m × 30 m × 40 m. How many cube-shaped crates, each with a side length of 2 m, can be stored in the warehouse?

(6) The dimensions of a cuboid-shaped tank are shown below:

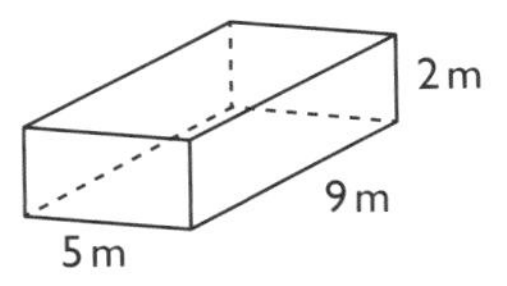

The tank is filled with water at a rate of 50 litres per minute. How long does it take to fill the tank?

Note: 1 litre = 1,000 cm^3

(7) The surface areas of 3 adjacent faces of a cuboid are 600 cm^2, 200 cm^2 and 300 cm^2. What is the capacity of the cuboid?

Note: 1 litre = 1,000 cm^3

litres

(8) The length of a cuboid is 20 cm. Its height and width are equal. What is the sum of its height and width if its capacity is 3.92 litres?

Note: 1 litre = 1,000 cm^3

☐☐ cm

Use the diagram below to help you answer questions 9 and 10.

Shapes A, B and C are each made from identical smaller cubes.

Shape A Shape B Shape C

(9) How many times greater is the volume of Shape C than the volume of Shape A?

☐☐

(10) If the volume of Shape C is 13,824 cm^3, what is the side length of each small cube?

☐ cm

Test 17: 2D Shapes

INSTRUCTIONS

You have 10 minutes to complete the following section.
You have 10 questions to complete within the time given.

Circle the letter beside the correct answer.

EXAMPLE

How many lines of symmetry does a regular octagon have?

A 5
B 6
C 7
(D) 8
E 9

The answer is **D**.
A regular octagon has 8 equal sides and 8 lines of symmetry.

(1) The area of a square hall is 64 m^2. What is the ratio of the width of the hall to its perimeter?

A 1:2
B 4:1
C 2:1
D 3:2
E 1:4

(2) Which statement about the shape below is TRUE?

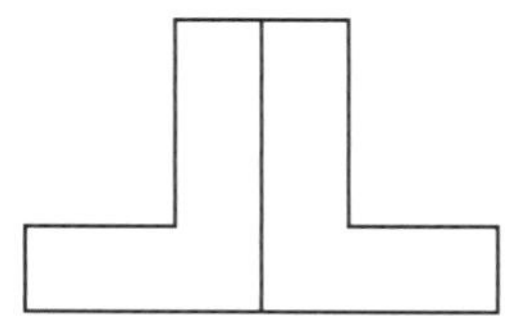

A The shape is a hexagon.
B The shape has no lines of symmetry.
C The shape has 2 lines of symmetry.
D The shape consists of 2 hexagons combined to form another hexagon.
E The shape consists of 2 hexagons combined to form an octagon.

3. The figure below consists of 2 identical regular hexagons. How many lines of symmetry does the figure have?

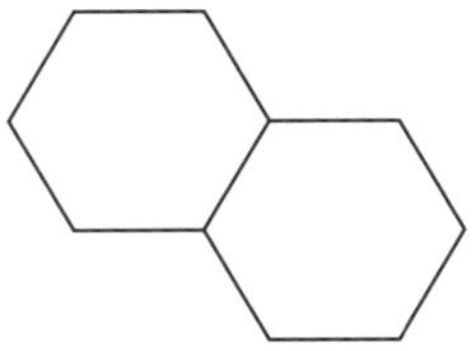

A 1
B 2
C 3
D 0
E 4

4. What is the order of rotational symmetry of the shape below?

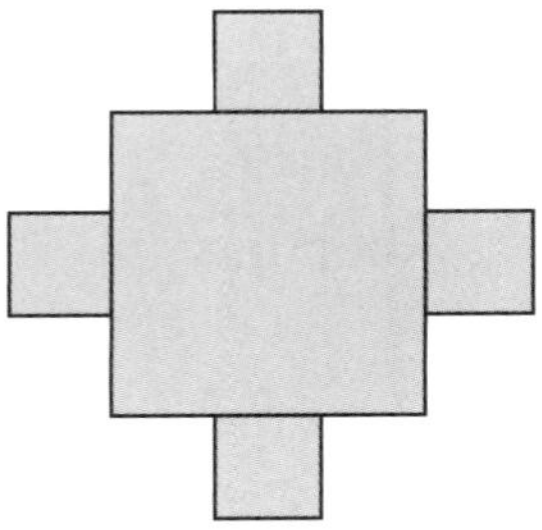

A 1
B 4
C 8
D 0
E 2

5. How many letters in the word **JOSEPH** have at least 2 lines of symmetry?

A 0
B 4
C 2
D 3
E 1

Use the diagram below to help you answer questions 6 and 7.

3 identical circles are enclosed inside a rectangle as shown below.

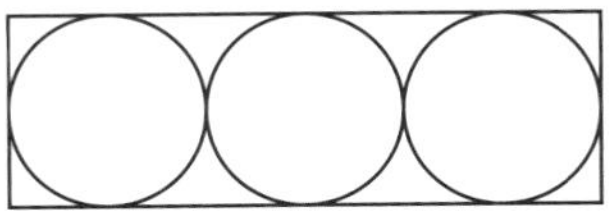

6. If the radius of each of the circles is 5 cm, what is the area of the rectangle?

A 300 cm^2
B 25 cm^2
C 75 cm^2
D 100 cm^2
E 200 cm^2

7. If the diameter of each circle is d, what is the relation between the length of the rectangle and d?

A Length of the rectangle = $6d$
B Length of the rectangle = d
C Length of the rectangle = $2d$
D Length of the rectangle = $3d$
E None of the above

Questions continue on next page

8. Zara describes a quadrilateral with the following properties:
   - Diagonally opposite angles are equal.
   - All sides are of equal length.
   - Opposite sides are parallel.
   - The diagonals bisect each other at 90°.

   What shape is she describing?

   A Rectangle
   B Kite
   C Trapezium
   D Triangle
   E Rhombus

HELPFUL TIP: *Remember that the answer has to have* ***all*** *of the described properties.*

Use the diagram below to help you answer questions 9 and 10.

Figure E consists of 3 concentric squares. Each square has a side length that is $\frac{5}{8}$ of the side length of the square immediately outside it. The outermost square has a perimeter of 128 cm.

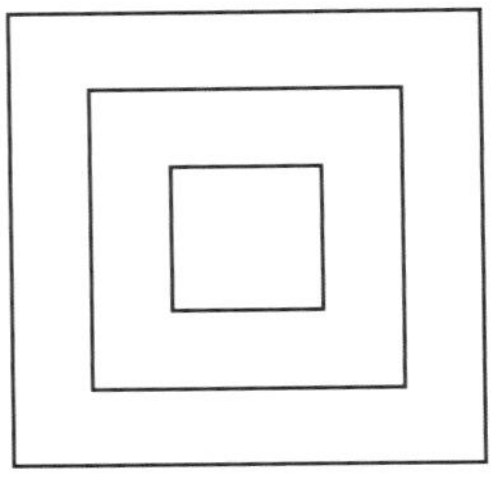

Figure E

9. What is the perimeter of the innermost square?

   A 50 cm
   B 17 cm
   C 16 cm
   D 80 cm
   E 4 cm

10. What is the area of the middle square in mm^2?

   A 4,000 mm^2
   B 40,000 mm^2
   C 400,000 mm^2
   D 400 mm^2
   E 40 mm^2

Test 18: Square and Cube Numbers

INSTRUCTIONS

You have 10 minutes to complete the following section.
You have 10 questions to complete within the time given.

Circle the letter beside the correct answer.

EXAMPLE

What is the difference between the square of 5 and the cube of 5?

A 5
B 25
C 125
D 75
(E) 100

The answer is **E**.

$5^2 = 5 \times 5 = 25$

$5^3 = 5 \times 5 \times 5 = 125$

$125 - 25 = 100$

(1) Which of these numbers is both a square number and a cube number?

A 8
B 1,000
C 121
D 216
E 64

(2) What is the cube root of 0.125?

A 0.05
B 0.25
C 0.5
D 1.25
E 0.55

Questions continue on next page

(3) How many numbers in the table below have at least 2 square numbers as factors?

100	18
200	75
160	81

A 6
B 3
C 1
D 2
E 4

(4) The difference between 19^2 and 6^3 is a multiple of which of these numbers?

A 3
B 2
C 4
D 5
E 7

(5) Brad divides 867 by x. His answer is a square number. What is the value of x?

A 4
B 6
C 3
D 2
E 5

(6) 3 calculations are shown below:

$11^2 = 121$

$101^2 = 10,201$

$1,001^2 = 1,002,001$

What is $10,001^2$?

A 1,000,002,000,001
B 10,000,020,001
C 10,002,001
D 1,000,020,001
E 100,020,001

HELPFUL TIP: *Look for a pattern in the listed calculations to help you.*

7. Which of these statements is TRUE?

A A number that ends with zero is always a square number.
B Square numbers never end with 6.
C No 2-digit square number ends in 3.
D The number of digits in a square number is always odd.
E None of the above.

8. Battalions of soldiers are asked to stand in rows such that each row has as many soldiers as there are rows. Which of the battalions shown below cannot form this arrangement?

	Number of Soldiers
Battalion 1	3,600
Battalion 2	4,800
Battalion 3	10,000
Battalion 4	160,000
Battalion 5	841

A Battalion 1
B Battalion 2
C Battalion 3
D Battalion 4
E Battalion 5

9. There are 2 groups, each with the same number of members. Each member contributes as many pounds as the number of members in the group. If the total money collected by the 2 groups is £288, how many members are there in each group?

A 12
B 21
C 9
D 13
E 15

10. Rebecca squares all the numbers in the table below and writes down her answers. How many of the numbers that Rebecca writes down are also cube numbers?

16	4	1	12	9	6	8

A 0
B 7
C 2
D 4
E 1

Test 19: Sequences

INSTRUCTIONS

You have 10 minutes to complete the following section.
You have 10 questions to complete within the time given.

Circle the letter beside the correct answer.

EXAMPLE

Which number comes next in this sequence?

45 50 55 60 65 ?

A 65
B 60
(C) 70
D 75
E 50

The answer is **C**.
Each term is 5 greater than the previous term
Next term = 65 + 5 = 70

(1) The first 4 terms in a sequence are as follows:

32, 27, 22, 17

Which term will be the 1st negative number in the sequence?

A 6th term
B 7th term
C 10th term
D 8th term
E 4th term

(2) The nth term of a sequence is given by the expression ($n^2 + 2n$). What is the 11th term of the sequence?

A 131
B 122
C 143
D 169
E 112

(3) The first 3 terms in a sequence are shown below:

$1 + a$, $1 + 2a$, $1 + 3a$

If the 4th term in the sequence is 25, what is the 1st term in the sequence?

A 6
B 7
C 8
D 9
E 3

(4) What is the next term in this sequence?

1, 25, 81, 169, ?

A 231
B 289
C 409
D 217
E 181

(5) A sequence of shapes is formed such that each shape in the sequence is twice as long as the previous shape. The width of each shape in the sequence is equal. The first 3 shapes in the sequence are shown below:

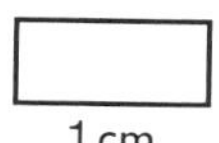

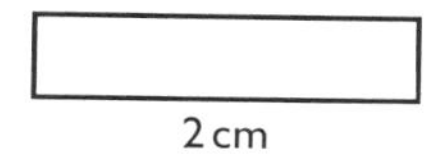

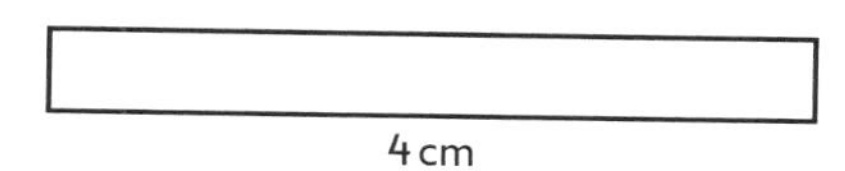

What is the area of the 5th shape if the area of the 2nd shape is 2 cm^2?

A 12 cm^2
B 32 cm^2
C 8 cm^2
D 16 cm^2
E 28 cm^2

(6) Which expression shows the nth term of the sequence below?

2, 6, 12, 20, 30

A $2n$
B $2n + 2$
C $n \times (n + 1)$
D $n + 1$
E $4n$

Questions continue on next page

7) Which number does not belong to the sequence below?

13, 18, 23, 28, 33 …

A 108
B 203
C 511
D 318
E 208

8) The nth term of a sequence is given by the expression $4n^2$. Which of these numbers belongs to this sequence?

A 144
B 180
C 260
D 150
E 104

Use the sequence below to help you answer questions 9 and 10.

9) These are the first 3 terms in a sequence:

Which term in the sequence will consist of 42 dots?

A 13th term
B 12th term
C 4th term
D 10th term
E 9th term

10) How many dots will there be in the 25th term in this sequence?

A 105
B 106
C 98
D 112
E 108

HELPFUL TIP: *Consider if there is a way to work this out without calculating the number of dots in each term from the 1st to the 25th.*

Test 20: Factors and Multiples

INSTRUCTIONS

You have 10 minutes to complete the following section.
You have 10 questions to complete within the time given.

Circle the letter beside the correct answer.

EXAMPLE

What is the sum of the factors of 8?

(A) 15
B 64
C 14
D 4
E 42

The answer is **A**.
Factors of 8 = 1, 8, 2 and 4
Sum = 1 + 8 + 2 + 4 = 15

(1) Some of the factors of 18 are shown below. What is the product of the factors of 18 that are not shown below?

3		1
	9	
18		

A 14
B 16
C 54
D 12
E 6

(2) What is half of the lowest common multiple of 15, 4 and 6?

A 12
B 40
C 30
D 20
E 68

Questions continue on next page

(3) 3 different alerts sound after every 32 seconds, 12 seconds and 80 seconds respectively. If they sound simultaneously at 8:19 p.m., at what time will they next sound simultaneously?

A 9:13 p.m.
B 8:27 p.m.
C 8:21 p.m.
D 8:24 p.m.
E 8:28 p.m.

HELPFUL TIP: *Think about how your understanding of Lowest Common Multiples can help you to answer this question.*

(4) What is the highest common factor of 240 and 320?

A 10
B 20
C 40
D 80
E 120

(5) This diagram shows the step lengths of 3 friends.

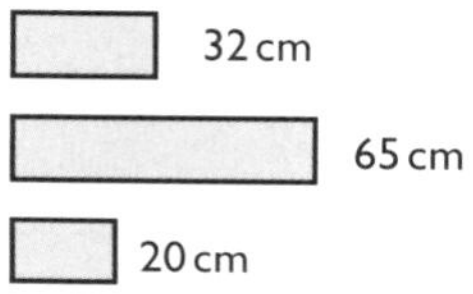

32 cm
65 cm
20 cm

What is the shortest possible distance they walk if they each cover exactly the same distance in full steps?

A 20.8 m
B 120 m
C 13 m
D 20.08 m
E 1 m

(6) Noah packs pasta into 100 g bags. Sam packs pasta into 200 g bags. Ella packs pasta into 250 g bags. Noah, Sam and Ella are each given the same weight of pasta and they each manage to pack it exactly. What weight of pasta were they each given?

A 1 kg
B 275 g
C 2.1 kg
D 800 g
E 675 g

7. The harvest from Hewitt's Farm is shown below.

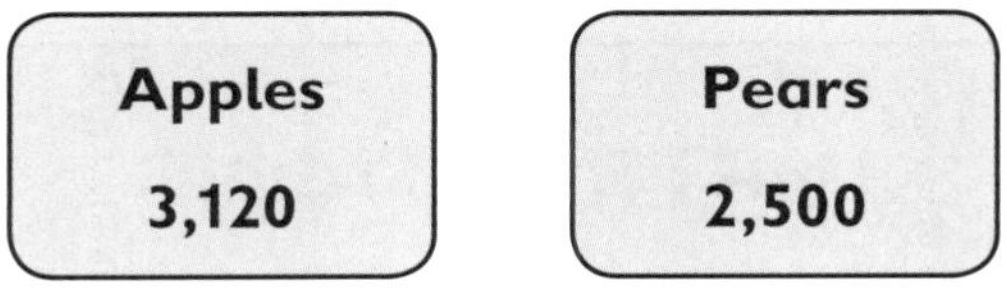

Apples	Pears
3,120	2,500

Kate divides the apples into equal piles, each containing X apples.

Jane divides the pears into equal piles, each containing X pears.

What is the greatest possible value of X?

A 20
B 25
C 60
D 48
E 10

8. The length and width of a rectangular room are shown below.

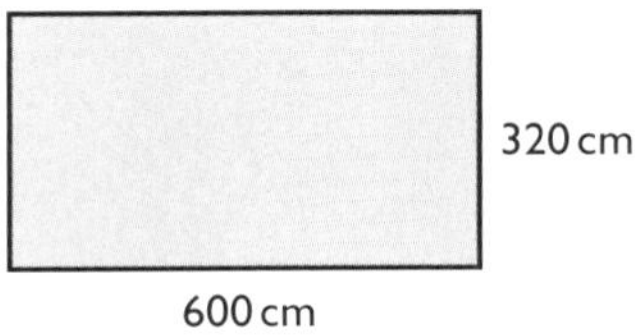

320 cm
600 cm

If the room is paved completely with a whole number of square tiles, what is the greatest possible side length of each tile?

A 3 m
B 250 cm
C 5 m
D 0.4 m
E 10 m

9. What is the largest number that will divide into 2,007, 2,403 and 1,035?

A 8
B 3
C 6
D 11
E 9

10. Which number will divide into 223, 246 and 181 leaving remainders of 3, 2 and 1 respectively?

A 4
B 3
C 2
D 1
E 8

Test 21: 3D Shapes

INSTRUCTIONS

You have 9 minutes to complete the following section.
You have 9 questions to complete within the time given.

Circle the letter beside the correct answer.

EXAMPLE

How many faces does a cube have?

A 5
(B) 6
C 7
D 8
E 10

The answer is **B**.
A cube has 6 faces.

(1) What is the sum of the number of faces of all 3 of these shapes?

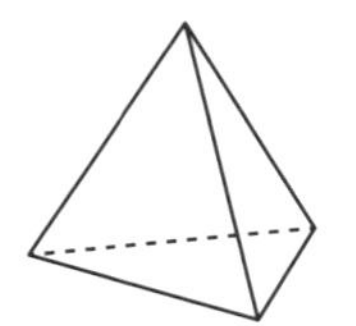
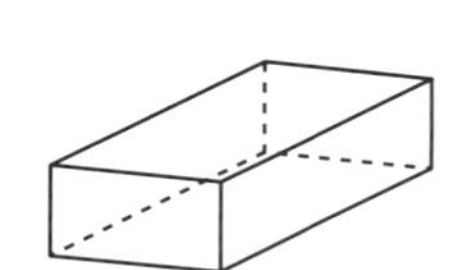
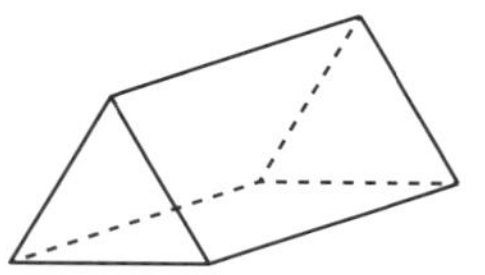

A 12
B 13
C 11
D 15
E 12

(2) Which of the following shapes has exactly half as many faces as edges?

A Cube
B Tetrahedron
C Sphere
D Cylinder
E Square-based pyramid

Use the diagram below to help you answer questions 3 and 4.

The 3D shape below is formed from identical cubes, each with a side length of 2 cm.

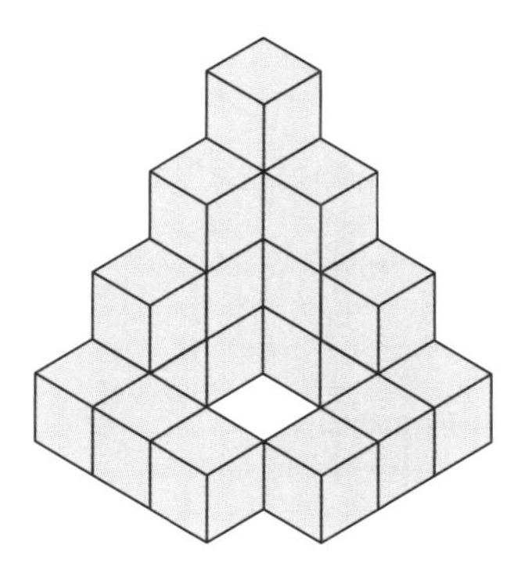

(3) What is the total volume of the shape?

A 80 cm^3
B 40 cm^3
C 160 cm^3
D 180 cm^3
E 200 cm^3

(4) How many cube faces are visible from the top-down bird's eye view of the shape?

A 6
B 8
C 10
D 11
E 16

---

(5) This net can be folded into a 3D shape. What is the name of the shape?

A Hexagonal prism
B Tetrahedron
C Hexagon-based pyramid
D Cuboid
E None of the above

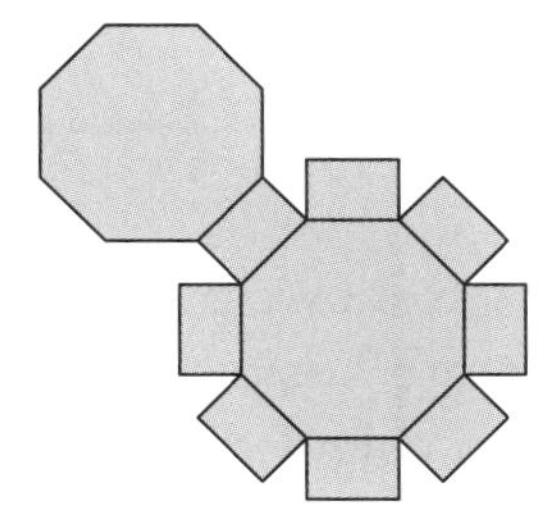

(6) A cuboid-shaped storage crate has the dimensions 3 m × 7 m × 5 m. If a football kit occupies 0.5 m^3 of space, what is the maximum number of kits that can be stored inside the crate?

A 300
B 215
C 310
D 210
E 200

Questions continue on next page

Use the diagram below to help you answer questions 7 and 8.

Figure D is formed from 20 identical small cubes.

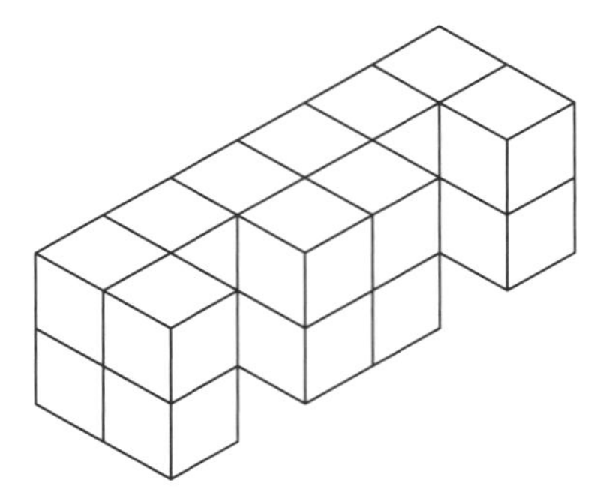

Figure D

(7) If the total volume of Figure D is 540 cm^3, what is the surface area of each of the cubes?

A 27 cm^2
B 9 cm^2
C 3 cm^2
D 36 cm^2
E 54 cm^2

(8) If Figure D is completely painted red, what fraction of the total number of small cube faces are painted red?

A $\frac{1}{2}$
B $\frac{1}{3}$
C $\frac{35}{120}$
D $\frac{80}{90}$
E $\frac{4}{5}$

---

(9) Figure E shows the top view of a cube-shaped crate that contains cylindrical cans. The crate has a side length of 10 cm and the total surface area of the 2 circular faces of each cylinder is 9.8 cm^2.

What is the total surface area of the circular faces shown in Figure E?

A 196 cm^2
B 156.8 cm^2
C 9.8 cm^2
D 78.4 cm^2
E 100 cm^2

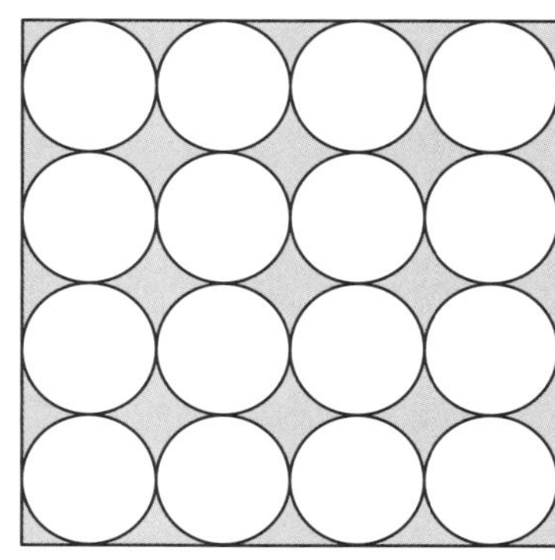

Figure E

Test 22: Place Value and Ordering

INSTRUCTIONS

You have 10 minutes to complete the following section.
You have 10 questions to complete within the time given.

Circle the letter beside the correct answer.

EXAMPLE

What is the value of the 8 in the number 78,321?

A 8
B 80
C 800
D 8,000
E 80,000

The answer is **D**.
8 is in the 'thousands' column
8 × 1,000 = 8,000

(1) What is the sum of the 3rd largest and the 2nd smallest numbers in the table below?

10,098	10,908	10,809	1,089	1,189	10,009

A 11,271
B 11,287
C 11,008
D 1,098
E 11,789

(2) Which of these numbers is closest to 0.03?

A 0.031
B 0.0031
C 0.31
D 0.0301
E 0.0295

Questions continue on next page

③ Alesha has driven 14,886 miles in her car since she bought it. She rounds this number to the nearest ten thousand. What is the difference between the rounded distance and the actual distance travelled?

A 15,115 miles
B 114 miles
C 4,886 miles
D 10,000 miles
E 4,000 miles

Use the number line below to help you answer questions 4 and 5.

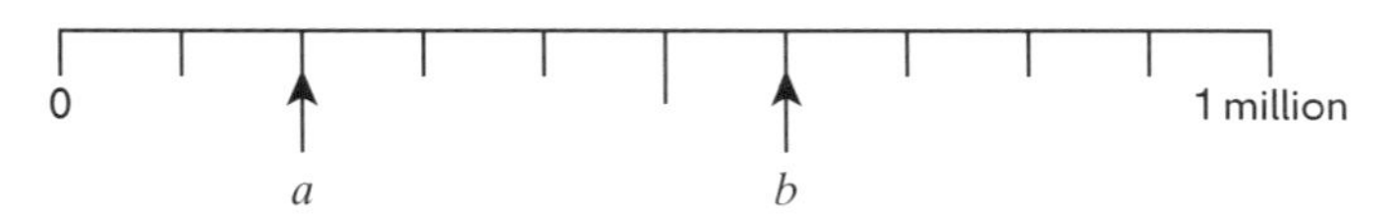

④ What is the value of $(a + b)$?

A 8,000,000
B 600,000
C 1,000,000
D 5,000,000
E 800,000

⑤ What is the place value of '4' in the number exactly midway between a and b?

A 1,000
B 10,000
C 100,000
D 10
E 100

⑥ A number has 6 digits. It is an odd multiple of 35. It rounds to 140,040 when rounded to the nearest 10. What is the value of the digit in the tens column of the number?

A 2
B 5
C 6
D 4
E 3

7. The price of a house was £395,000 last year. The price has increased by $\frac{1}{10}$ this year. What is the place value of the digit '3' in the increased price?

A Hundred
B Thousand
C Ten thousand
D Ten
E Million

8. A museum has as many artefacts as the difference between the largest 4-digit even whole number and the smallest 3-digit odd whole number. How many artefacts does the museum have?

A 9,897
B 9,998
C 9,898
D 9,899
E 998

9. There are 191 houses with the postcode SW67. The houses are numbered consecutively from 1 to 191. The houses numbered from 78 to 101 will be demolished to build a leisure centre. How many houses will remain in the postcode?

A 167
B 176
C 166
D 168
E 171

10. John adds all the numbers shown on the clock face below.

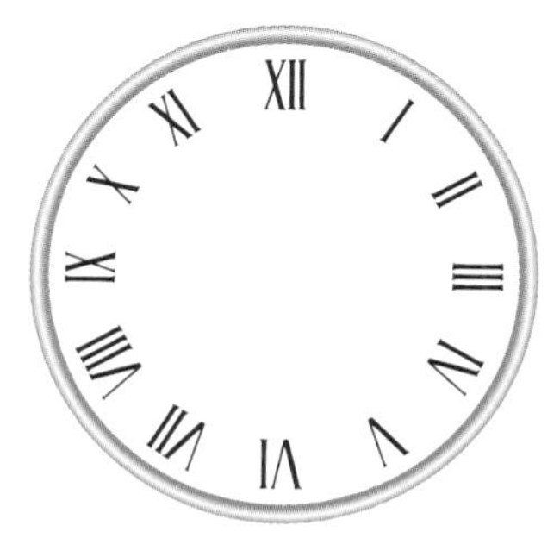

Which of these Roman numerals represents John's answer?

A LXVIII
B CXXVIII
C LXXVIII
D VIIVIII
E XCIV

Notes

Answers

Key abbreviations: °C: degrees centigrade, cm: centimetre, d.p.: decimal place, g: gram, HCF: Highest Common Factor, kg: kilogram, km: kilometre, l: litre, LCM: lowest common multiple, m: metre, mins: minutes, ml: millilitre, mm: millimetre, r: remainder

Test 1: Algebra

Q1 C
Gabby had $4n$ pencils; Gabby's sister had $2n$ pencils; total number of pencils = $4n + 2n = 6n$
$6n \div 2 = 3n$

Q2 D
$6x + 6x + 8 = 40.4$; $12x = 40.4 - 8 = 32.4$
$x = 32.4 \div 12 = 2.7$

Q3 A
12 years ago, Debbie was 28 years old.
12 years ago, Mandy was $28s$ years old.
Today, Mandy is $(28s + 12)$ years old.

Q4 D
Cost of pizza =
total cost – cost of 4 drinks = $4t - d$

Q5 B
$\frac{b^2}{2a} = -5^2 \div (2 \times 4) = 25 \div 8 = 3.125$

Q6 C
$5p - 4 = 61$; $5p = 61 + 4 = 65$
$p = 65 \div 5 = 13$; $p + 7 = 13 + 7 = 20$

Q7 B
You can use trial and error with answer choices:
If Bella is 15 then Abi must be 9.
Bella's age in 9 years = 15 + 9 = 24
Abi's age in 9 years = 9 + 9 = 18
Ratio = 18:24 = 3:4

Q8 B
Total cost in pence =
$12c + 15d + 5c + 10c + 8d = 27c + 23d$
Total cost in pounds = $\frac{(27c + 23d)}{100}$

Q9 A
Let capacity = C; $\frac{3}{8} C = x$

$C = x \div \frac{3}{8} = x \times \frac{8}{3} = \frac{8x}{3}$

Q10 D
Amount of water remaining = original amount of water minus amount of water Leon drinks in ml = $x - 30$

Test 2: Decimals

Q1 A
Worked example: Let X = 90
90 ÷ 10 = 9; 90 ÷ 0.1 = 900; 900 ÷ 9 = 100

Q2 C
24 × 1.96 = 47.04
47.04 rounded to the nearest whole number is 47

Q3 C
Cost of electricity = 20 p × 10.1 = 202 p
Cost of gas = 10 p × 15.2 = 152 p
Total = 152 p + 202 p = 354 p
Cost after reduction =
354 p × 50% = 177 p = £1.77

Q4 C
Cost of pencils = 2,000 × £0.21 = £420
Percentage = (£420 ÷ £3,000) × 100 = 14%

Q5 D
Total baggage weight =
12.9 kg + 7.5 kg + (5 × 0.6 kg) = 23.4 kg
Excess = 23.4 kg – 23 kg = 0.4 kg = 400 g

Q6 B
Total number of bags =
120.4 kg ÷ 0.2 kg = 602
Number of bags sold = $\frac{1}{7} \times 602 = 86$
Money received = 86 × £3.50 = £301

Q7 B
Weight of 4 boxes of sweets =
8.8 kg ÷ 2 = 4.4 kg; weight of 1 box of sweets = 4.4 kg ÷ 4 = 1.1 kg

Q8 A
Average = 876.2 miles ÷ 6.5 days = 134.8 miles
134.8 miles rounded to the nearest mile is 135 miles

Q9 C
Total rainfall = 9.59 cm × 4 = 38.36 cm
February rainfall = 38.36 cm –
10.2 cm – 8.6 cm – 11.56 cm = 8 cm

Q10 E
Weight of 1 candy = 4,400 g ÷ 200 = 22 g
Weight of Pile B =
120 × 22 g = 2,640 g = 2.64 kg

Test 3: Fractions

Q1 60%
Grade A minimum mark = $\frac{3}{4} \times 25 = 18.75$
Percentage awarded Grade A =
$\frac{3}{5} \times 100 = 60\%$

Q2 68.6 kg
Increase in weight = $\frac{2}{5} \times 49$ kg = 19.6 kg
Total weight = 49 kg + 19.6 kg = 68.6 kg

Q3 **0.25 m**

After 1st bounce = $\frac{1}{4}$ × 4 m = 1 m

After 2nd bounce = $\frac{1}{4}$ × 1 m = 0.25 m

Q4 **4**

After 3rd bounce =

$\frac{1}{4}$ × 0.25 m = 0.0625 m = 6.25 cm

After 4th bounce = $\frac{1}{4}$ × 6.25 cm = 1.5625 cm

Q5 **12.5 cm**

Length of matchstick = $\frac{5}{12}$ × 30 cm = 12.5 cm

Q6 **3 minutes**

$\frac{3}{5}$ of an hour = 36 minutes; total travel time = 36 minutes + 7 minutes = 43 minutes

Minutes late = 43 minutes – 40 minutes = 3 minutes

Q7 $\frac{1}{6}$

Number of faces painted = 8

Total number of faces = 48

Fraction of surface area painted = $\frac{8}{48} = \frac{1}{6}$

Q8 **6**

Total number of blocks = 16 × 3 = 48

Blocks made of steel = $\frac{1}{8}$ × 48 = 6

Q9 **25%**

Fraction voting for Mr Widmore =

$1 - \frac{2}{5} - \frac{7}{20} = \frac{20}{20} - \frac{8}{20} - \frac{7}{20} = \frac{5}{20} = \frac{1}{4}$

$\frac{1}{4}$ = 25%

Q10 **£65.20**

Number of cupcakes sold = $\frac{2}{7}$ × 322 = 92

92 × £0.60 = £55.20; number of cookies sold = $\frac{1}{2}$ × 200 = 100; 100 × £0.90 = £90

Remaining money = £55.20 + £90 – £80 = £65.20

Test 4: Angles

Q1 **D**

Triangle ABC is isosceles.

Angle B = 180° – 112° = 68°

y = 180° – 68° – 68° = 44°

Q2 **B**

$\frac{\text{Sum of interior angles of a scalene triangle} = 180°\ \text{Sum of interior angles of a quadrilateral}}{3}$

= 360° ÷ 3 = 120°

An exterior angle of an equilateral triangle × 1.5 = 120° × 1.5 = 180°

Angles on a straight line = 180°

Sum of interior angles of a triangle = 180°

Q3 **D**

Internal angle of an equilateral triangle = 180° ÷ 3 = 60°; a = 360° – 60° = 300°

Q4 **A**

q = 180° – 65° – 52° = 63°

Q5 **D**

k is one angle in an isosceles triangle.

Size of each of the other angles in this isosceles triangle = 180° – interior angle of a regular pentagon = 180° – 108° = 72°

k = 180° – 72° – 72° = 36°

Q6 **B**

90° right from east is south.

90° right from south is west.

Q7 **E**

Turning east to southwest in a clockwise direction is equivalent to 1.5 right-angled turns; 1.5 × 90° = 135°

Q8 **B**

Size of angle between adjacent numbers on clock face = 360° ÷ 12 = 30°

Reflex angle between hands at 3:45 p.m. is equal to 6.75 times this angle.

30° × 6.75 = 202.5°

Q9 **A**

Interior angle of a regular pentagon = 108°

a is one angle in an isosceles triangle.

a = (180° – 108°) ÷ 2 = 36°

Q10 **C**

a is equal in size to the angle adjacent to b.

Therefore, (a + b) = the interior angle of a regular pentagon.

Alternative proof: b, b and (108° – 2a) are 3 angles in a triangle

So b = (180° – (108° – 2a)) ÷ 2 = (180° – 36°) ÷ 2 = 144° ÷ 2 = 72°

So a + b = 36° + 72° = 108° = interior angle of a regular pentagon

Test 5: Money

Q1 **122**

£300 ÷ £2.45 = 122.448979…

So 122 is the greatest number of tiles possible.

Q2 **£192.80**

Repair charges = (£35.75 × 2) + (£22.50 × 2) + £13 + (3 × £6.10) = £147.80

Labour costs = 1.5 hours × £30 = £45

Total cost = £147.80 + £45 = £192.80

Q3 **£20.50**

Weekly earnings = (25 × £11.50) + (9 × £13) = £404.50

Difference = £425 – £404.50 = £20.50

Q4 **£1.30**
Cost of 1 litre = £19 ÷ 15 = £1.26666...
£1.2666... rounded to nearest 10 p is £1.30

Q5 **Coffee**
Total paid = £10 – £4.55 = £5.45
3 colas + 2 teas + 1 coffee =
(3 x £0.65) + (2 x £1.10) + £1.30 = £5.45
So highest priced item is coffee.

Q6 **£1,300**
Profit per pizza = £4.75 – £1.50 = £3.25
Total profit = 4 × 100 × £3.25 = £1,300

Q7 **£2.40**
Cost of 24 plants at Greenhouse =
£5.10 × 4 = £20.40; cost of 24 plants at
Growell = £3.80 × 6 = £22.80
Saving = £22.80 – £20.40 = £2.40

Q8 **£111.50**
Cost with meter = ((365 days × 3 hours) × £0.30) + £100 = £428.50
Saving = £540 – £428.50 = £111.50

Q9 **£647.50**
Cost with meter = ((365 days × 5 hours) × £0.30) + £100 = £647.50

Q10 **25**
Cost per dictionary = £3 × 1.15 = £3.45
Dictionaries sold = £86.25 ÷ £3.45 = 25

Test 6: Prime Numbers

Q1 **A**
97 ÷ 2 = 48.5

Q2 **D**
Prime numbers less than 30: 2, 3, 5, 7, 11, 13, 17, 19, 23, 29

Q3 **C**

Q4 **D**
51 – 33 = 18

Q5 **A**
179 + 347 = 526

Q6 **A**
Prime factors of 30: 2, 3, 5

Q7 **B**
A prime number's only factors are itself and 1.

Q8 **B**
$2^2 \times 3^2 = 4 \times 9 = 36$

Q9 **E**
2 × 6, 3 × 4

Q10 **C**
111 ÷ 3 = 37; so 111 is not prime

Test 7: Percentage

Q1 **C**
85% of cost = £34; 5% of cost = £34 ÷ 17 = £2
100% of cost = £2 × 20 = £40

Q2 **A**
15% = 9; 5% = 3; 100% = 3 x 20 = 60
Number who chose Honey Loops =
60 – 12 – 9 – 15 – 11 = 13

Q3 **E**
Cost of dress = £320 × 1.17 = £374.40
Cost of ring = £290 × 1.18 = £342.20
Difference = £374.40 – £342.20 = £32.20

Q4 **E**
Non-fiction books = 20% × 220 = 44
Biographies = $\frac{5}{11} \times 44 = 20$

Q5 **A**
New length = 10 cm × 1.2 = 12 cm
New width = 4 cm × 1.2 = 4.8 cm
Area = 12 cm × 4.8 cm = 57.6 cm^2

Q6 **C**
$\frac{3}{10} \times a = b;\ a = b \div \frac{3}{10}$
$a = b \times \frac{10}{3};\ a = \frac{10b}{3}$

Q7 **A**
Worked example:
Let x be £100
July price = £100 × 0.8 = £80
August price = £80 × 1.25 = £100

Q8 **E**
Percentage = $\frac{162°}{360°} \times 100\% = 45\%$

Q9 **D**
Number of Italians = 45% × 7,500 = 3,375
Number of French = (7,500 – 3,375) ÷ 3 = 1,375

Q10 **C**
Worked example: Let Frank's income be 5 so Alfred's income is 10. 5 must double to become 10 so Frank's income must increase by 100%.

Test 8: Probability

Q1 **1**
All months in 2020 will have more than 28 days (it is a leap year so there will be 29 days in February).

Q2 $\mathbf{\frac{11}{18}}$
Number of letters = 18
Number of consonants = 11
Probability = $\frac{11}{18}$

Q3 **0.02**
Probability = 10% × $\frac{1}{5}$ = 0.1 × 0.2 = 0.02

Q4 **0.4**
2 shapes have no lines of symmetry.
Probability = $\frac{2}{5}$ = 0.4

Q5 **3**

Fraction of green or blue ducks = $1 - \frac{5}{14} = \frac{9}{14}$

Twice as many blue ducks as green ducks.

Probability of choosing a blue duck = $\frac{6}{14}$

Probability of choosing a green duck = $\frac{3}{14}$

So the smallest possible number of green ducks is 3.

Q6 **0.42**

Total number of students = 21 + 25 + 14 = 60

Total number of students preferring drama = 25

Probability = $\frac{25}{60}$ = 0.4166...

0.4166 ... rounded to 2 d.p. is 0.42

Q7 **0.24**

Probability = $\frac{4}{5} \times \frac{3}{10} = \frac{12}{50} = 0.24$

Q8 **0.75**

Number of red sections = 6

Probability = $\frac{6}{8}$ = 0.75

Q9 **0.2**

2 out of 10 students have black hair

2 ÷ 10 = 0.2

Q10 $\mathbf{\frac{1}{6}}$

Number of bags weighing less than 4.8 kg = 1

Probability = $\frac{1}{6}$

Test 9: Ratio

Q1 **D**

Number of parts = 1 + 5 = 6

Simon's share = $\frac{5}{6} \times$ £30 = £25

Money left = 80% × £25 = £20

Q2 **B**

Let width = X; length = $2X$

$X + X + 2X + 2X = 96$ cm

$6X = 96$ cm; $X = 96$ cm ÷ 6 = 16 cm

Length = 2 × 16 cm = 32 cm

Q3 **C**

Total area of carpet = $\frac{2}{6} \times 360 \times 20\ m^2 = 2{,}400\ m^2$

Q4 **A**

Length of longer piece = $3y - 12$

Ratio = $(3y - 12)$:12

Q5 **B**

140 ÷ 35 = 4; Time required = 4 × 12 minutes = 48 minutes

Q6 **E**

Number of hours required = 4 × 15 × 8 = 480 hours

Number of days now required = 480 ÷ (10 × 12) = 480 ÷ 120 = 4

Q7 **A**

Number of grey tiles per slab = 10

Number of grey tiles used = 20 × 10 = 200

Area of 1 tile = 20 cm × 20 cm = 400 cm^2

Total area = 400 cm^2 × 200 = 80,000 cm^2

80,000 cm^2 = 8 m^2

Q8 **C**

Total number of white tiles used = 20 × 8 = 160

Total cost of tiles = (160 × £0.50) + (200 × £2.10) = £500

Q9 **E**

240:183 = 80:61

Q10 **A**

Flour required for 19 pies = (210 g ÷ 6) × 19 = 665 g

250 g × 2 = 500g; 250 g × 3 = 750 g

So 3 packets will be needed.

Test 10: Perimeter

Q1 **C**

Perimeter of field = 24 km ÷ 6 = 4 km

Width of field = (4 km − 1.5 km − 1.5 km) ÷ 2 = 0.5 km = 500 m

Q2 **D**

Area of Cloth A = 100 cm × 400 cm = 40,000 cm^2; area of Cloth B = 210 cm × 210 cm = 44,100 cm^2

44,100 cm^2 – 40,000 cm^2 = 4,100 cm^2

Q3 **E**

Perimeter = 220 m × 4 = 880 m = 0.88 km

Cost = £65 × 0.88 = £57.20

Q4 **B**

Length of shortest side = (2,150 m – 400 m – 400 m – 400 m – 400 m) ÷ 2 = 275 m

Q5 **B**

Q6 **C**

Ratio is 6:3 = 2:1

Q7 **A**

224 mm + 108 mm + 150 mm = 482 mm = 48.2 cm; 1.85 cm × 3 = 5.55 cm

155 mm + 155 mm + 30 mm = 340 mm = 34 cm

0.75 cm + 1 cm + 1.25 cm = 3 cm

0.5 cm + 1 cm + 1.5 cm = 3 cm

Q8 **B**

Trial and error method: white square side length = 120 cm ÷ 4 = 30 cm

Area = 30 cm × 30 cm = 900 cm^2

Grey square side length = 60 cm ÷ 4 = 15 cm

Area = 15 cm × 15 cm = 225 cm^2

Uncovered area = 900 cm^2 – 225 cm^2 = 675 cm^2

Q9 **E**

x and y are equal

Q10 **C**

Distance in widths = 4 × 40 m = 160 m
Distance in lengths = 160 m × 2 = 320 m
Number of lengths = 320 m ÷ 80 m = 4

Test 11: Graphs and Charts

Q1 **D**

Acute angle is less than 90°
90° is equivalent to 25% on the chart
4 responses are less than 25%

Q2 **C**

Size of smallest garden =
117 m^2 – 32 m^2 = 85 m^2; Since this value does not already exist on the table, it must be the area of the garden of House 6.

Q3 **D**

This is the best way to represent 2 sets of comparative data.

Q4 **C**

Angle = 35% × 360° = 126°

Q5 **A**

Range = 13 – 1 = 12

Q6 **B**

Difference on Day 4 = 15 – 13 = 2

Q7 **D**

Weekday hours = 12 hours × 5 = 60 hours
Weekend hours = 9 hours + 6 hours = 15 hours
Difference = 60 hours – 15 hours = 45 hours

Q8 **A**

Earnings per Sunday = 6 hours × £11 = £66
Earnings per Monday = 12 hours × £9.20 = £110.40; Earnings per fortnight =
2 × (£66 + £110.40) = £352.80
£352.80 rounded to the nearest £100 is £400

Q9 **C**

The first part of the graph consists of one straight line, slanted upwards, showing that the child was moving at a constant speed. When the line is horizontal, time is passing but there is no change in distance, meaning there is no movement.

Test 12: Tables

Q1 $\frac{2}{9}$

Number of shapes = 9
Number of shapes with exactly 4 lines of symmetry = 2; probability = $\frac{2}{9}$

Q2 **0**

All the black shapes have at least 1 line of symmetry.

Q3 **09:38**

He must arrive at York Racecourse by 10:03 at the latest so he must take the 09:38 bus.

Q4 **6:55 p.m.**

If he takes the 6:55 p.m. bus, he will arrive at 7:07 p.m.

Q5 **13:41**

Total time until end of lunch break = 15 mins + 22 mins + 144 mins + 40 mins = 221 mins
221 mins after 10:00 a.m. = 1:41 p.m. = 13:41

Q6 **1 hour 48 minutes**

Sports roundup = 1.8 × 60 minutes =
108 minutes = 1 hour 48 minutes

Q7 **19 hours**

Hours worked on Monday = 6 hours 25 mins
Hours worked on Tuesday = 5 hours 50 mins
Hours worked on Wednesday = 6 hours 45 mins
Total hours worked = 19 hours

Q8 **6**

Prime factors of 900: 2, 2, 3, 3, 5, 5
Prime factors of 250: 2, 5, 5, 5
HCF = 2 × 5 × 5 = 50
Prime factors of 25: 5, 5
Prime factors of 60: 2, 2, 3, 5
LCM = 2 × 2 × 3 × 5 × 5 = 300
$\frac{b}{a}$ = 300 ÷ 50 = 6

Q9 **x = 8, y = 14**

Prime factors of 8: 2, 2, 2
Prime factors of 14: 2, 7
LCM = 2 × 2 × 2 × 7 = 56

Q10 **09:38**

Time taken by Bus A = 94 mins
Time taken by Bus B = 99 mins
Total time taken = 112 mins × 3 = 336 mins
Time taken by Bus C = 336 mins – 94 mins – 99 mins = 143 mins
143 mins after 07:15 is 09:38

Test 13: Time

Q1 **C**

Time taken on phonics = 35 mins × 5 days × 8 weeks = 1,400 mins
1,400 mins = $23\frac{1}{3}$ hours = 23 hours (rounded to the nearest hour)

Q2 **B**

Time in Beijing is 7 hours ahead of the time in London. So actual time in Beijing is 22:10. Grandad's clock is 25 minutes slow.

Q3 **E**

Film finishes at 12:35 a.m. Beijing time.
This is 5:35 p.m. in London; 5:35 p.m. = 17:35

Q4 **B**
Stefan → 312 seconds
Sia → 306 seconds
Hannah → 360 seconds
Jenny → 330 seconds
Meera → 512 seconds

Q5 **E**
Mean time = (312 + 306 + 360 + 330 + 512) ÷ 5 = 364 seconds

Q6 **D**
Lesson 2 for Year 1 ends 10 minutes after it ends for Year 6.
10 minutes after 12:00 p.m. is 12:10 p.m.

Q7 **A**
Lesson 4 for Year 1 starts 20 minutes after it starts for Year 6.
20 minutes after 2:10 p.m. is 2:30 p.m.

Q8 **D**
Her brother's day is 20 minutes longer due to 10 minutes added to the Break and 10 minutes added to Lunch.

Test 14: Weights and Measures

Q1 **D**
Sum = 11 m + 140 mm + 2,100 mm + 2,000 mm + 95 mm = 4,346 mm

Q2 **B**
40% of 3,000 ml = 1,200 ml
Number of cups = 1,200 ml ÷ 50 ml = 24

Q3 **C**
Number of coins Lauren has = 348 g ÷ 12 g = 29; value of Lauren's coins = 29 × £2 = £58
Value of coins required to make £70 = £70 – £58 = £12; number of coins needed to make £70 = £12 ÷ £2 = 6

Q4 **B**
Number of coins in line = £70 ÷ £2 = 35
Length of line = 35 × 28.4 mm = 994 mm = 99.4 cm

Q5 **A**
1.44 km = 1,440 m = 144,000 cm
Distance on map = 144,000 cm ÷ 1,200 = 120 cm

Q6 **C**
0.08 tonnes = 80 kg = 80,000 grams

Q7 **C**
Side length of square = 1 cm
Area of square = 1 cm × 1 cm = 1 cm^2

Q8 **A**
Yvonne bought 0.065 kg of cinnamon.
Cost of cinnamon = 0.065 × £25 = £1.625 = £1.60 (rounded to nearest 10 p)

Q9 **D**
Greatest possible weight = 29,499 g
Smallest possible weight = 28,500 g
Difference = 29,499 g – 28,500 g = 999 g

Q10 **C**
Monkey's weight 5 months ago = 44 kg
Amount of weight monkey loses = 20% × 44 kg = 8.8 kg
Amount of weight monkey loses per month = 8.8 kg ÷ 5 = 1.76 kg

Test 15: Area

Q1 **15 cm**
Base × height × $\frac{1}{2}$ = area
Height × 7.1 cm = 106.5 cm^2
Height = 106.5 cm^2 ÷ 7.1 cm = 15 cm

Q2 **2.5 m**
Width = area ÷ length = 112.5 m^2 ÷ 45 m = 2.5 m

Q3 **99%**
Total area = 200 m × 45 m = 9,000 m^2
Percentage covered by concrete = (112.5 m^2 ÷ 9000 m^2) × 100 = 1.25%
Percentage covered by grass = 100% – 1.25% = 98.75%
99% is the closest answer option.

Q4 **50%**
Area = length (L) × width (W) = LW
New area $\frac{L}{2}$ × 3W = 1.5 LW = 50% more than LW

Q5 **37 cm**
Base × height × $\frac{1}{2}$ = area
Base × 10 cm × $\frac{1}{2}$ = 75 cm^2
Base = 75 cm^2 ÷ 5 cm = 15 cm
Perimeter = 15 cm + 11 cm + 11 cm = 37 cm

Q6 **36 cm^2**
Area of each rectangle = 2 cm × 1.5 cm = 3 cm^2
Area of H-shape = 3 cm^2 × 12 = 36 cm^2

Q7 **£237.50**
Perimeter = 120 ÷ 6 = 20 m; side length = 20 m ÷ 4 = 5 m; area = 5 m × 5 m = 25 m^2
Cost = 25 × £9.50 = £237.50

Q8 **240**
Area of frame = (30 cm × 30 cm) – (20 cm × 15 cm) = 600 cm^2
Number of buttons = (600 cm^2 ÷ 10 cm^2) × 4 = 60 × 4 = 240

Q9 **4,800 cm^2**
Enlarged area of picture = 60 cm × 80 cm = 4,800 cm^2

Q10 **25:4**
Side length ratio is $(10 \div 4):(4 \div 4) = 2.5:1$
Area ratio is $2.5^2:1^2 = 6.25:1 = 25:4$

Test 16: Volume

Q1 **2,400 cm^3**
Total volume =
20×8 cm $\times 5$ cm $\times 3$ cm $= 2{,}400$ cm^3

Q2 **£1,562.50**
Volume of pit = 5 m × 5 m × 5 m = 125 m^3
Cost = 125 m^3 × £12.50 = £1,562.50

Q3 **6 cm**
Total number of cubes = $11 \times 4 = 44$
Volume of 1 cube = 9,504 cm^3 ÷ 44 = 216 cm^3
Side length of 1 cube = $\sqrt[3]{216} = 6$ cm

Q4 $\frac{5}{11}$
Number of cubes in lowest layer = 20
Fraction of total = $\frac{20}{44} = \frac{5}{11}$

Q5 **9,000**
Warehouse volume = 60 m × 30 m × 40 m = 72,000 m^3
Volume of 1 crate = 2 m × 2 m × 2 m = 8 m^3
Number of crates = 72,000 m^3 ÷ 8 m^3 = 9,000

Q6 **30 hours**
Volume of tank = 500 cm × 900 cm × 200 cm = 90,000,000 cm^3
Volume of tank in litres = 90,000,000 cm^3 ÷ 1,000 cm^3 = 90,000 litres
Time taken = 90,000 litres ÷ 50 litres = 1,800 minutes = 30 hours

Q7 **6 litres**
Let dimensions of cuboid = x, y and z
$xy = 600$ cm^2; $xz = 300$ cm^2; $yz = 200$ cm^2
Then use trial and error to work out that dimensions of cuboid are 30 cm × 20 cm × 10 cm; volume of cuboid = 30 cm × 20 cm × 10 cm = 6,000 cm^3; capacity = 6,000 cm^3 ÷ 1,000 cm^3 = 6 litres

Q8 **28 cm**
Volume of cuboid = 3.92 litres × 1,000 cm^3 = 3,920 cm^3; height × width = 3,920 cm^3 ÷ 20 cm = 196 cm^2; height = width = $\sqrt{196}$ = 14 cm
Height + width = 28 cm

Q9 **64**
Number of small cubes in Shape A = $2 \times 2 \times 2 = 8$; number of small cubes in Shape C = $8 \times 8 \times 8 = 512$; number of times C is greater than A = $512 \div 8 = 64$

Q10 **3 cm**
Volume of each small cube = 13,824 cm^3 ÷ 512 = 27 cm^3
Side length of each small cube = $\sqrt[3]{27} = 3$ cm

Test 17: 2D Shapes

Q1 **E**
Width = $\sqrt{64\text{ m}^2}$ = 8 m
Perimeter = 8 m × 4 = 32 m
Ratio = 8:32 = 1:4
Check: The perimeter of a square will always be four times its width so the ratio of width to perimeter will always be 1:4

Q2 **E**

Q3 **B**

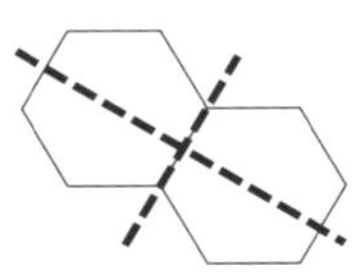

Q4 **B**

Q5 **C**
O and H

Q6 **A**
Length of rectangle = 6 × radius = 6 × 5 cm = 30 cm; width of rectangle = 2 × radius = 2 × 5 cm = 10 cm
Area = 30 cm × 10 cm = 300 cm^2

Q7 **D**
Diameter = 2 × radius
Length of rectangle = 6 × radius = 3 × diameter = $3d$

Q8 **E**

Q9 **A**
Side length of outer square = 128 cm ÷ 4 = 32 cm; side length of middle square = 32 cm × $\frac{5}{8}$ = 20 cm; side length of inner square = 20 cm × $\frac{5}{8}$ = 12.5 cm; perimeter of inner square = 12.5 cm × 4 = 50 cm

Q10 **B**
Side length of middle square = 20 cm = 200 mm
Area = 200 mm × 200 mm = 40,000 mm^2

Test 18: Square and Cube Numbers

Q1 **E**
$8^2 = 64$; $4^3 = 64$

Q2 **C**
$0.5 \times 0.5 \times 0.5 = 0.125$

Q3 **A**
All of them have 1 as a factor and at least 1 other square number.

Q4 **D**
$19^2 - 6^3 = 145$; $145 \div 5 = 29$

Q5 **C**
$864 \div 3 = 289$; $\sqrt{289} = 17$

Q6 E

The pattern indicates that there will be 3 zeros between each of the other value digits in the number.

Q7 C

2-digit square numbers: 16, 25, 36, 49, 64, 81

Q8 B

4,800 is not a square number.

Q9 A

Money collected by each group = £288 ÷ 2 = £144

Members in each group = $\sqrt{144}$ = 12

Q10 C

1 × 1 = 1; 1 × 1 × 1 = 1

8 × 8 = 64; 4 × 4 × 4 = 64

Test 19: Sequences

Q1 D

Each term is 5 less than the previous term.

32, 27, 22, 17, 12, 7, 2, –3

Q2 C

11th term = $11^2 + (2 \times 11) = 121 + 22 = 143$

Q3 B

4th term in the sequence is $(1 + 4a)$

$1 + 4a = 25$; $4a = 24$; $a = 6$

1st term is $(1 + a)$; $1 + a = 1 + 6 = 7$

Q4 B

Sequence is $1^2, 5^2, 9^2, 13^2$… so the number being squared in each term is 4 greater than the number being squared in the previous term.

So next term is $17^2 = 289$

Q5 D

Width of each shape = 2 cm^2 ÷ 2 cm = 1 cm

Length of 5th shape = 4 cm × 2 × 2 = 16 cm

Area of 5th shape = 16 cm × 1 cm = 16 cm^2

Q6 C

Check: when $n = 4$, the term is 20

$n \times (n + 1) = 4 \times 5 = 20$

Q7 C

All numbers in the sequence are 3 greater than a multiple of 5.

Q8 A

$4n^2 = 144$; $n^2 = 36$; $n = 6$

Q9 E

10 dots, 14 dots, 18 dots…; each term has 4 more dots than the previous term.

10, 14, 18, 22, 26, 30, 34, 38, 42

Q10 B

Check: 10, 14, 18, 22, 26, 30, 34, 38, 42, 46, 50, 54, 58, 62, 66, 70, 74, 78, 82, 86, 90, 94, 98, 102, 106

Test 20: Factors and Multiples

Q1 D

Factors of 18 not shown: 2, 6; 2 × 6 = 12

Q2 C

Prime factors of 15: 3, 5

Prime factors of 4: 2, 2

Prime factors of 6: 2, 3

LCM = 3 × 5 × 2 × 2 = 60; 60 ÷ 2 = 30

Q3 B

Prime factors of 32: 2, 2, 2, 2, 2

Prime factors of 12: 2, 2, 3

Prime factors of 80: 2, 2, 2, 2, 5

LCM = 2 × 2 × 2 × 2 × 2 × 3 × 5 = 480

480 seconds = 8 minutes

8 minutes after 8:19 p.m. is 8:27 p.m.

Q4 D

Prime factors of 240: 2, 2, 2, 2, 3, 5

Prime factors of 320: 2, 2, 2, 2, 2, 2, 5

HCF = 2 × 2 × 2 × 2 × 5 = 80

Q5 A

Prime factors of 32: 2, 2, 2, 2, 2

Prime factors of 65: 5, 13

Prime factors of 20: 2, 2, 5

LCM = 2 × 2 × 2 × 2 × 2 × 5 × 13 = 2,080 cm = 20.8 m

Q6 A

Prime factors of 100: 2, 2, 5, 5

Prime factors of 200: 2, 2, 2, 5, 5

Prime factors of 250: 2, 5, 5, 5

LCM = 2 × 2 × 2 × 5 × 5 × 5 = 1,000 g = 1 kg

Q7 A

Trial and error method using answer choices:

3,120 ÷ 20 = 156; 2,500 ÷ 20 =125

20 is the greatest possible answer choice that divides into both numbers.

Q8 D

Trial and error method using answer choices:

320 cm ÷ 40 cm = 8; 600 cm ÷ 40 cm = 15

0.4 m is the largest answer choice that divides into both the length and the width exactly.

Q9 E

Trial and error method using answer choices:

2,007 ÷ 9 = 223; 2,403 ÷ 9 = 267

1,035 ÷ 9 = 115; 9 is the largest answer choice that divides into all 3 numbers.

Q10 A

Trial and error method using answer choices:

223 ÷ 4 = 55 r 3; 246 ÷ 4 = 61 r 2

181 ÷ 4 = 45 r 1; 4 is the answer choice that fits the stated requirements.

Test 21: 3D Shapes

Q1 **D**

Triangle-based pyramid has 4 faces; cuboid has 6 faces; triangular prism has 5 faces.
$4 + 6 + 5 = 15$

Q2 **A**

Number of faces of a cube = 6
Number of edges of a cube = 12
$12 \div 6 = 2$

Q3 **C**

Number of cubes = 20; volume of 1 cube = $2\text{ cm} \times 2\text{ cm} \times 2\text{ cm} = 8\text{ cm}^3$
Total volume = $8\text{ cm} \times 20 = 160\text{ cm}^3$

Q4 **D**

Q5 **E**

The shape is actually an octagonal prism.

Q6 **D**

Volume of crate = $3\text{ m} \times 7\text{ m} \times 5\text{ m} = 105\text{ m}^3$
Volume of 1 football kit = 0.5 m^3
Number of kits = $105\text{ m}^3 \div 0.5\text{ m}^3 = 210$

Q7 **E**

Volume of 1 cube = $540\text{ cm}^3 \div 20 = 27\text{ cm}^3$
Side length of 1 cube = $\sqrt[3]{27\text{ cm}^3} = 3\text{ cm}$
Area of 1 face of 1 cube = $3\text{ cm} \times 3\text{ cm} = 9\text{ cm}^2$
Surface area of 1 cube = $9\text{ cm}^2 \times 6 = 54\text{ cm}^2$

Q8 **A**

Total number of faces painted red = 60
Total number of faces = $20 \times 6 = 120$
Fraction painted red = $\frac{60}{120} = \frac{1}{2}$

Q9 **D**

Surface area of 1 circular face = $9.8\text{ cm}^2 \div 2 = 4.9\text{ cm}^2$; total surface area shown = $4.9\text{ cm}^2 \times 16 = 78.4\text{ cm}^2$

Test 22: Place Value and Ordering

Q1 **B**

$10{,}098 + 1{,}189 = 11{,}287$

Q2 **D**

$0.031 - 0.03 = 0.001$; $0.03 - 0.0031 = 0.0269$; $0.31 - 0.03 = 0.28$; $0.0301 - 0.03 = 0.0001$; $0.03 - 0.0295 = 0.0005$; 0.0001 is the smallest difference so 0.0301 is the closest to 0.03

Q3 **C**

Rounded distance = 10,000 miles
Difference = 14,886 miles – 10,000 miles = 4,886 miles

Q4 **E**

$200{,}000 + 600{,}000 = 800{,}000$

Q5 **C**

Halfway between 200,000 and 600,000 is 400,000

Q6 **E**

$140{,}040 \div 35 = 4{,}001$ r 5; $4{,}001 \times 35 = 140{,}035$
140,035 rounded to the nearest 10 is 140,040
So the number is 140,035

Q7 **C**

Increased price = $1.1 \times$ £395,000 = £434,500

Q8 **A**

Number of artefacts = $9{,}998 - 101 = 9{,}897$

Q9 **A**

No. 78 to No. 101 $\rightarrow$ 24 houses in total
Number of remaining houses = $191 - 24 = 167$

Q10 **C**

Sum = $1 + 2 + 3 + 4 + 5 + 6 + 7 + 8 + 9 + 10 + 11 + 12 = 78$; LXXVIII = $50 + 10 + 10 + 5 + 1 + 1 + 1 = 78$

Score Sheet

Below is a score sheet to track your results over multiple attempts. One mark is available for each question in the tests.

Test	Pages	Date of first attempt	Score	Date of second attempt	Score	Date of third attempt	Score
Test 1: Algebra	4–6						
Test 2: Decimals	7–9						
Test 3: Fractions	10–12						
Test 4: Angles	13–15						
Test 5: Money	16–18						
Test 6: Prime Numbers	19–21						
Test 7: Percentage	22–24						
Test 8: Probability	25–27						
Test 9: Ratio	28–30						
Test 10: Perimeter	31–33						
Test 11: Graphs and Charts	34–36						
Test 12: Tables	37–39						
Test 13: Time	40–42						
Test 14: Weights and Measures	43–45						
Test 15: Area	46–48						
Test 16: Volume	49–51						
Test 17: 2D Shapes	52–54						
Test 18: Square and Cube Numbers	55–57						
Test 19: Sequences	58–60						
Test 20: Factors and Multiples	61–63						
Test 21: 3D Shapes	64–66						
Test 22: Place Value and Ordering	67–69						